Advances in Microelectronics

Advances in Microelectronics

Edited by **Eve Versuh**

WILLFORD PRESS

New York

Published by Willford Press,
118-35 Queens Blvd., Suite 400,
Forest Hills, NY 11375, USA
www.willfordpress.com

Advances in Microelectronics
Edited by Eve Versuh

International Standard Book Number: 978-1-68285-019-0 (Hardback)

Printed in the United States of America.

Contents

Preface

The main aim of this book is to educate learners and enhance their research focus by presenting diverse topics covering this vast field. This is an advanced book which compiles significant studies by distinguished experts. This book addresses successive solutions to the challenges arising in the area of application, along with it; the book provides scope for future developments.

Microelectronics is a branch of electronics which is of utmost importance in the present age where electronic devices are omnipresent. Discussed in this book are various concepts like miniaturization, applications of micromachines, microrobotics, micromachinery, microtransducers, etc. While understanding the long-term perspectives of the topics, the book makes an effort in highlighting their impact as a modern tool for the growth of the discipline. This book is a resource guide for professionals as well as students engaged in the field of microelectronics and allied subjects.

It was a great honour to edit this book, though there were challenges, as it involved a lot of communication and networking between me and the editorial team. However, the end result was this all-inclusive book covering diverse themes in the field.

Finally, it is important to acknowledge the efforts of the contributors for their excellent chapters, through which a wide variety of issues have been addressed. I would also like to thank my colleagues for their valuable feedback during the making of this book.

Editor

A Passive Pressure Sensor Fabricated by Post-Fire Metallization on Zirconia Ceramic for High-Temperature Applications

Tao Luo [1], Qiulin Tan [1,2,3,*], Liqiong Ding [1], Tanyong Wei [2], Chao Li [2], Chenyang Xue [1,2] and Jijun Xiong [1,2,*]

[1] Key Laboratory of Instrumentation Science and Dynamic Measurement (North University of China), Ministry of Education, North University of China, Taiyuan 030051, China;
E-Mails: 18935157540@163.com (T.L.); ding418liqiong@163.com (L.D.);
xuechenyang@nuc.edu.cn (C.X.)

[2] Science and Technology on Electronic Test and Measurement Laboratory, North University of China, Taiyuan 030051, China; E-Mails: wei418tany@163.com (T.W.); li418chao@163.com (C.L.)

[3] National Key Laboratory of Fundamental Science of Micro/Nano-Device and System Technology, Chongqing University, Chongqing 400044, China

* Authors to whom correspondence should be addressed; E-Mails: tanqiulin.99@163.com (Q.T.);
xiongjijun@nuc.edu.cn (J.X.)

External Editor: Nicolaas F. De Rooij

Abstract: A high-temperature pressure sensor realized by the post-fire metallization on zirconia ceramic is presented. The pressure signal can be read out wirelessly through the magnetic coupling between the reader antenna and the sensor due to that the sensor is equivalent to an inductive-capacitive (LC) resonance circuit which has a pressure-sensitive resonance frequency. Considering the excellent mechanical properties in high-temperature environment, multilayered zirconia ceramic tapes were used to fabricate the pressure-sensitive structure. Owing to its low resistivity, sliver paste was chosen to form the electrical circuit via post-fire metallization, thereby enhancing the quality factor compared to sensors fabricated by cofiring with a high-melting-point metal such as platinum, tungsten or manganese. The design, fabrication, and experiments are demonstrated and discussed in detail. Experimental results showed that the sensor can operate at 600 °C with quite good coupling. Furthermore, the average sensitivity is as high as 790 kHz/bar within the measurement range between 0 and 1 Bar.

Keywords: pressure sensor; high temperature; post-fire metallization; zirconia ceramic; LC resonance

---

1. Introduction

The monitoring and management of performance and system health of hypersonic aerial vehicles, jet engines, rockets, and so on, relies on the capability to detect a variety of parameters including pressure [1]. Usually, these applications are accompanied with high-temperature environments, which issue a great challenge to current commercial pressure sensors. For example, the pressure in a turbine engine needs to be measured from 300 °C to 1000 °C at different locations [2,3].

Harsh environments such as gas turbine engines, power plants, and material-processing systems generally have temperatures greater than 500 °C [4]. The typical temperature in these environments is beyond the working temperature of traditional silicon-based devices. For example, the working-temperature range of pressure sensors fabricated in silicon would be less than 200 °C because the electrical and mechanical properties of silicon deteriorate at higher temperature. Owing to temperature characteristics of the P-N junctions, the highest operational temperature of silicon-based electronics is limited to approximately 130 °C. When temperature exceeds 500 °C, the mechanical properties of silicon begin to degrade [5]. In addition, the electrical wires connecting the device to the electrical signal processing unit will overheats the processing unit causing the measurement system to fail quickly.

Researchers have developed various methods such as active cooling and thermal insulation packaging to extend the operation temperature of current pressure sensors [6]. However, experiments have shown that most traditional sensors can only be operated at approximately 300 °C, despite the well-designed package, owing to the limited temperature resistance of the kernel sensing unit. Additionally, those package strategies will cause the volume expansion of components, which will lead to poor installation adaptability. In addition to the packaging method, guide pipes have been used for pressure measurement in an engine running test, which usually cause signal distortion compared to the results obtained from *in situ* measurement. Therefore, further studies have focused on fabricating sensors by high-temperature-resistant materials such as ceramics [7–9], silicon on insulator (SOI) [10], and SiC [11–13]. Among these sensors, wireless passive pressure sensors realized by low-temperature cofired ceramic (LTCC) and high-temperature cofired ceramic (HTCC) technology have demonstrated great prospects owing to the combination of passive a telemetry readout method and temperature-resistant material properties.

In this paper, a wireless passive zirconia-based pressure sensor fabrication by the post-fire metallization is proposed. The sensor is fabricated by two main steps: sintering of the multilayered zirconia tapes and post-fire metallization of Ag pastes separately, which enhances the quality of the sensor compared to that of a sensor fabricated previously by cofiring with platinum pastes. Detailed fabrication processes are presented and the sensor is calibrated using an established calibration system from which an average sensitivity up to 790 kHz/bar is obtained. Further, the high-temperature characteristics of the sensor are investigated from room temperature to 600 °C. At 600 °C, the induced impedance phase dip of the reader antenna related to the proposed sensor is 50° which is quite considerable compared with 6° induced by the cofired sensor.

2. Sensor Design

A schematic of the sensor system is presented in Figure 1, from which it is clear that an LC resonance circuit formed by an invariable inductance coil and a variable capacitor is powered by a reader antenna. From the flow chart in Figure 1, the capacitance C_s increases owing to the reduction in the gap between the two capacitor electrodes if pressure is applied to the device, *i.e.*, the gap size of the cavity between the two membranes decreases. Therefore, the resonant frequency f_0 decreases, which can be coupled with the antenna port through the magnetic link M between the coils. In addition, the resonance information of the sensor can be extracted and tracked by measuring impedance parameters (phase, real part, magnitude, *etc.*) of the antenna.

A cross-sectional model of the sensor consisting of three ceramic tapes is shown in Figure 2. Before high-temperature sintering, the dimensions of the green tape are 42 mm × 42 mm, and the dimensions are approximately 35 mm × 35 mm after high-temperature sintering owing to shrinkage. The properties of the implemented zirconia ceramic are listed in Table 1. A sealed cavity is located in the middle of Layer 2, and a capacitor is electrically connected to the inductor, which is designed as a circular planar spiral coil (PSC) and placed on the top surface of Layer 3. The capacitor is a parallel-plate capacitor with circular electrodes in Layer 1 and Layer 3. The relevant geometrical parameters of the capacitor and inductor are listed in Table 2. The electrical connection between the bottom capacitor plates and the inductance coil is realized by a metalized via. Thus, a passive LC resonator is formed by these components. Furthermore, this type of electrical connection can also be achieved by coating the side wall of the sensor with Ag paste, thereby making the via hole unnecessary. However, a via is implemented in the proposed sensor instead of coating the side wall with Ag paste to achieve a more stable connection and to reduce the complexity of manual manipulation.

Figure 1. Measurement principle and structure diagram of the pressure sensor.

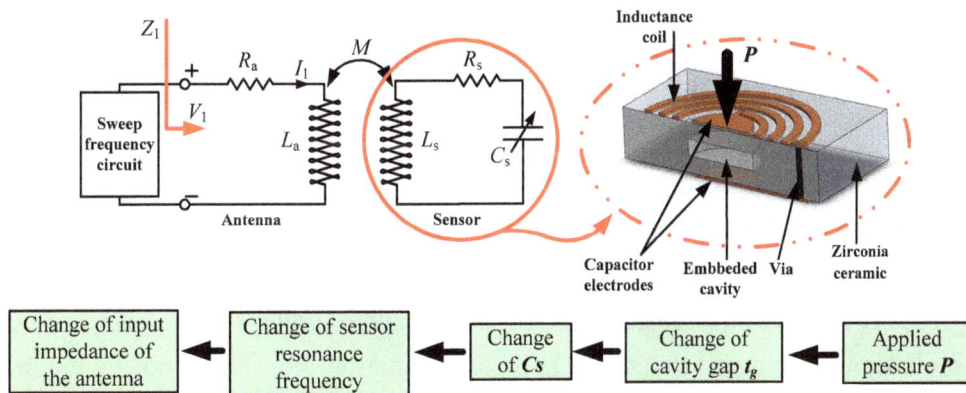

Figure 2. Cross-sectional model and layout of the sensor.

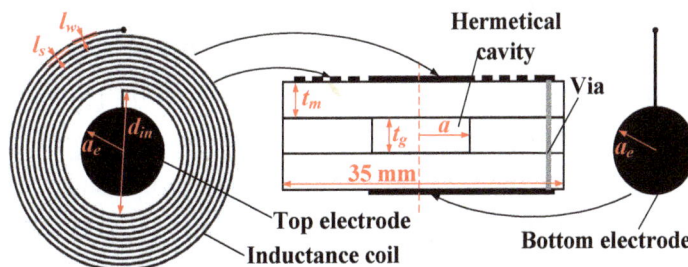

Table 1. Characteristics of the zirconia ceramic.

Quantity	Value
Young's modulus	210 GPa
Poisson's Ratio	0.24
Unfired thickness	125 μm
X,Y Shrinkage	17.3% ± 1%
Z Shrinkage	18.3% ± 1%

Table 2. Geometrical parameters in inductor and capacitor design.

Component	Symbol	Quantity	Value
Inductor	d_{in}	inner diameter of the inductance coils	14 mm
	l_w	width of the coils	200 μm
	l_s	spacing of the coils	400 μm
	n	number of the coils	10
Capacitor	a	radius of the cavity	3.6 mm
	a_e	radius of the electrode	3.8 mm
	t_g	cavity thickness	125 μm
	t_m	green-tape thickness	125 μm

3. Sensor Fabrication

The sensor was fabricated by separately sintering the ceramic substrate and metallic pattern. The fabrication process, as illustrated in Figure 3, can be roughly divided into punching, via filling, collating, lamination, high-temperature sintering, screen-printing, and-low temperature sintering. In order to provide clearer information about the fabrication process of the sensor, a comparison between the post-fire metallization method and the traditional method using the co-firing technique is also depicted in Figure 3.

Before punching, the ceramic tape was placed into a dry furnace preheated at 80 °C for 30 min. Then, a numerically controlled punching machine was used to drill a cavity hole in Layer 2 and via holes in all three layers. Next, via holes were filled in by the Pt paste to realize an electrical connection throughout the multilayered tapes. After via filling, Layers 1 and 2 were precisely stacked by the collating machine. Then, the stacked two-layer substrate was removed from the stacking machine, and a carbon membrane which has the same dimensions as the cavity and can volatilize within 600 °C, was placed into the cavity to support the pressure-sensitive membranes to avoid collapsing and cracking during the lamination process. It should be noted that the area of carbon membrane should not be much less than the area of the cavity because it will also lead to cracking in the edge of the cavity. Afterwards, Layer 3 was collated on top of Layer 2.

After stacking, lamination was performed, which is quite crucial for the quality of the buried cavity. The first step in the process of lamination is vacuum packaging of the ceramic tape, which ensured that the tapes were isolated from the water during the lamination. Then, the packaged ceramic tape was placed into the lamination machine at a pressure of 15 MPa and temperature of 75 °C for 20 min to make the multilayered tapes an integer. The lamination pressure and temperature should be controlled properly. That is, an exorbitant temperature and pressure will lead to the collapse of the sensitive membrane; conversely, an undertemperature or underpressure will lead to the delamination. After lamination, the 8-inch ceramic tape is cut into sensor sheets. Finally, the laminated substrate was sintered in a furnace at peak temperature of 1510 °C, as shown in Figure 4. During sintering process, the carbon membrane turns into CO_2 and volatilizes through apertures of the ceramic tapes because the tape is not air-tight when the organic compounds are baked off.

After sintering the zirconia substrate, the inductor and capacitor patterns were constructed by post-fire metallization with DuPont 6142D Ag paste. The capacitor electrode and inductor coil were printed on the surface of the zirconia substrate by a manual screen-printing platform, in which X and Y axes can be adjusted, as shown in Figure 5. There are several suction holes that can rigidly fix the sensor in place. After printing the electrical pattern, the semifinished sensor was placed into a baking furnace at 200 °C for 15 min to dry the paste. Then the sensor sample was placed into a seven-zone belt furnace to metalize the inductor and capacitor. The belt of the furnace operates at a speed of 100 mm/min in accordance with sintering data Table 3. The fabricated sensor is shown in Figure 6. It can be observed that the sacrificial layer has volatilized completely, and the flatness of the membrane is fairly good.

Figure 3. Fabrication process diagram of the sensor.

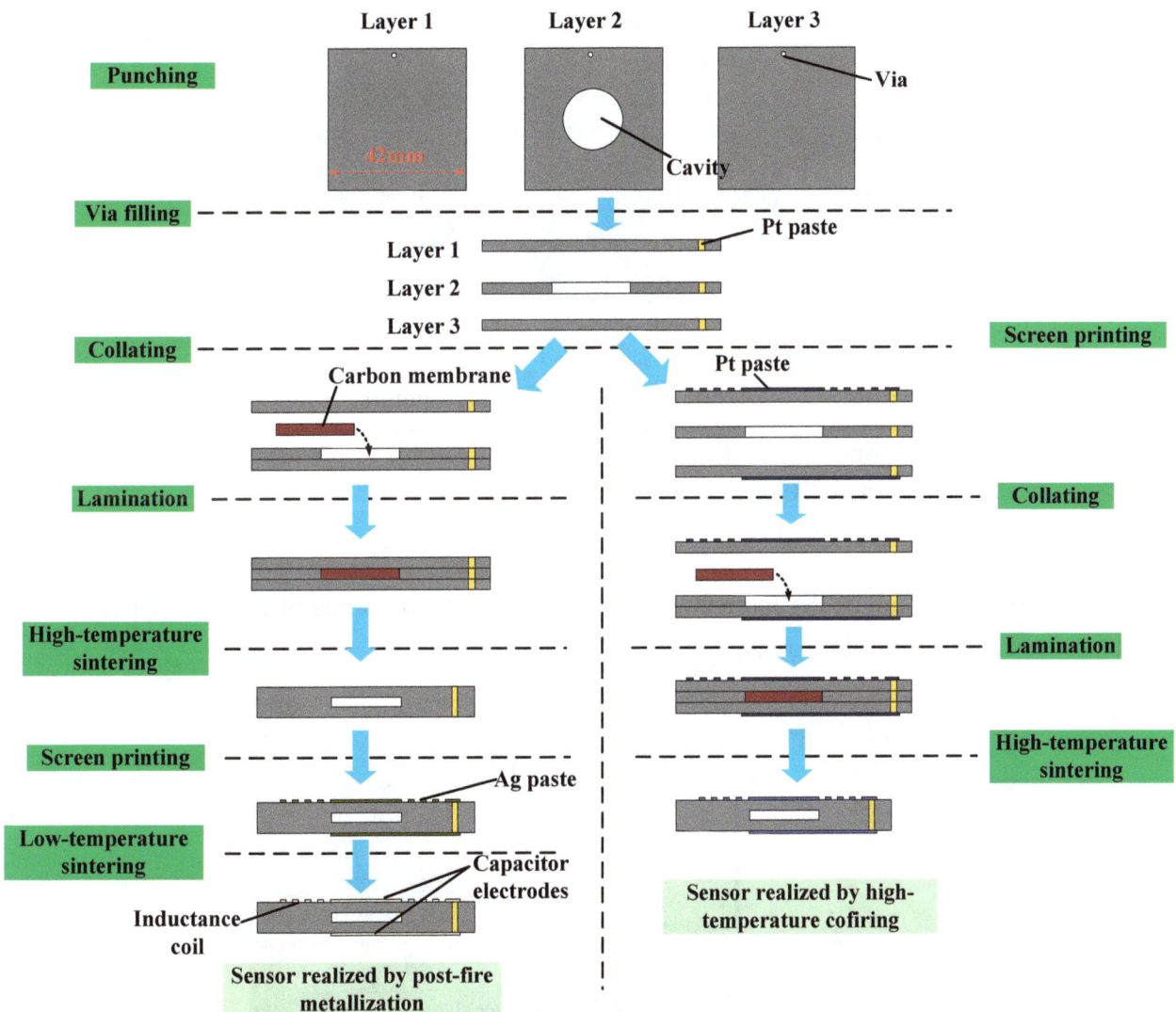

Figure 4. High-temperature sintering curve.

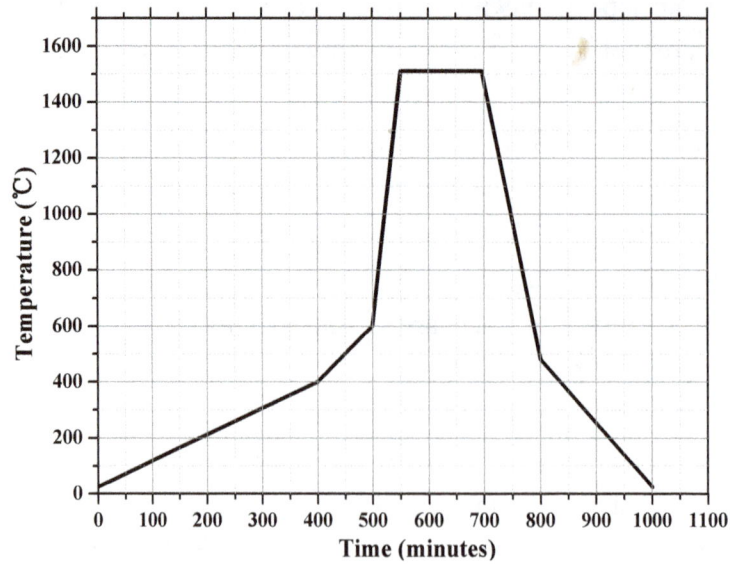

Figure 5. Manual screen-printing platform.

Table 3. Low-temperature sintering parameters.

Zone	1	2	3	4	5	6	7
Temperature (°C)	525	625	850	850	850	850	680
Speed (mm/min)	100	100	100	100	100	100	100

Figure 6. (**a**) Photograph of the fabricated sensor (**b**) photomicrographs of the inductance coil; (**c**) via; and (**d**) capacitor cavity.

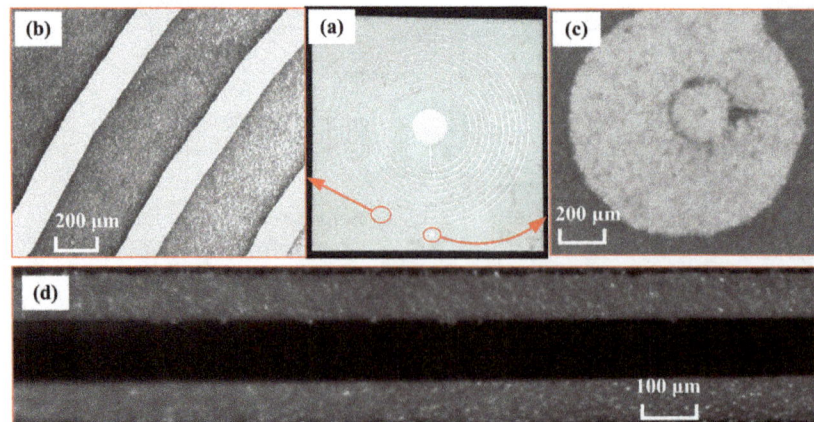

4. Experimental Section

As shown in Figure 7, the pressure calibration system consists of an Agilent E5061B (Agilent Technologies Inc., Santa Clara, CA, USA) Network Analyzer, a pressure vessel, a pressure controller, and a nitrogen tank. The reader antenna and sensor were placed into the pressure vessel, and the sensor was mounted within the near-field coupling distance (approximately 20 mm) of the antenna so that the sensor signal could be read out wirelessly. The distance between the antenna and the sensor has considerable influence on the wireless signal readout. With an increase in the coupling distance, the signal of sensor becomes weaker. However, a relatively better signal strength can be obtained when the self-resonance frequency of the antenna is closer to the resonance frequency of the sensor, even though the distance between them is constant. Therefore, a printed circuit board (PCB)-based planar spiral coil was used as a reader antenna, and a capacitor was added in series to make their resonance frequencies similar.

Figure 7. Sensor test setup for pressure measurement.

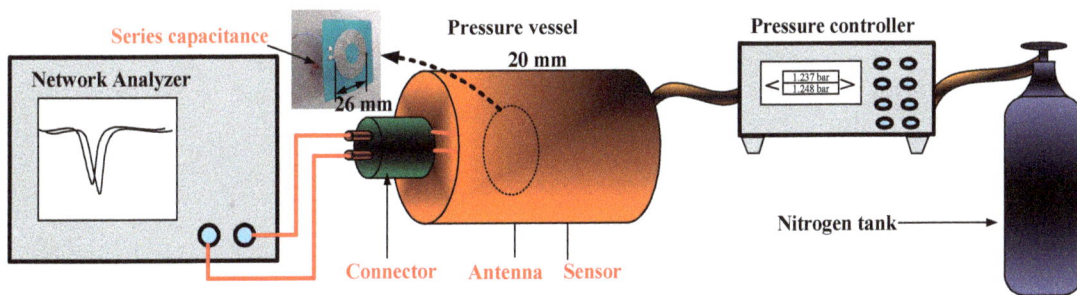

The test results are illustrated in Figure 8, from which it is clear that the sensitivity decreases as the applied pressure increase. Within the measurement range of 0–1 bar, the sensor exhibited a considerable average sensitivity of 790 kHz/bar. It should be noted that the linearity of the sensor deteriorates with an increase in the ambient pressure because the small deflection theory is not suitable over the entire measurement range. With an increase in pressure, large deformation occurs, which will result in nonlinear deformation of the sensitive membrane and eventually lead to a nonlinear pressure response from the sensor. Undoubtedly, the sensitive membrane has some predeflection due to the fabrication process, even though the carbon membrane was used to improve the flatness of the membrane, which can greatly impair the linearity of the sensor. In Figure 8, the measurement data points were fitted with the following quadratic function:

$$y = 22.924 - 1.469x + 0.683x^2 \tag{1}$$

where y is the frequency value corresponding to the negative peak point of the phase of the antenna, and x is the pressure applied to the surface of the sensor.

The temperature measurement setup is shown in Figure 9. A muffle furnace was used to provide an environment with precise temperature control, and a tungsten filament was used as a coil for the reader antenna owing to its better stability at high temperature. The test results show that the resonant frequency of sensor decreases as the temperature increase, as show in Figure 10. The nonlinear temperature response is mainly due to the inherent temperature-dependent dielectric constant of the zirconia ceramic. In actual application, a compensation scheme should be developed to compensate for the temperature drift so that the sensor can be used in different temperature environments. With regard to the dynamic range, the sensor proposed in this paper can operate within 1 bar at temperatures greater than 600 °C. The sensitivity and dynamic range for pressure measurement of the sensor can be extended by further adjusting the process parameters (sintering curve, lamination pressure, temperature, *etc.*) to improve the flatness of the pressure-sensitive membrane. The coupling effect, which can be reflected by the value of the phase dip, became worse when the temperature increased. Furthermore, the deteriorating coupling effect is caused by the decrease in the quality factor, which is mainly due to the increase in the series resistance and capacitance, as illustrated in Equation (2).

Figure 8. Measured frequency *versus* pressure.

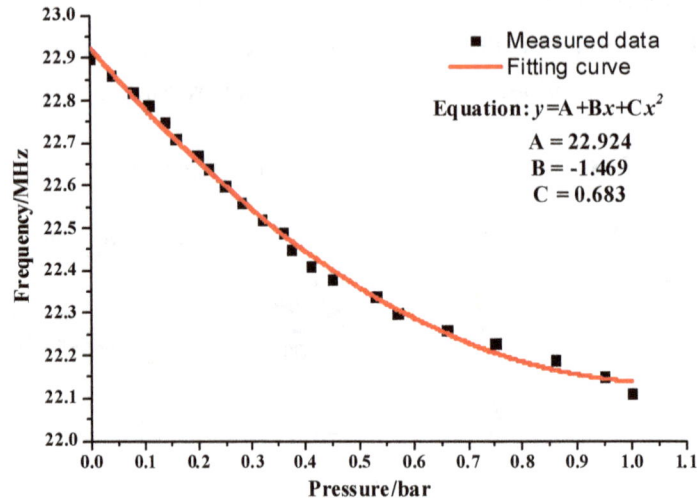

Figure 9. High-temperature testing system.

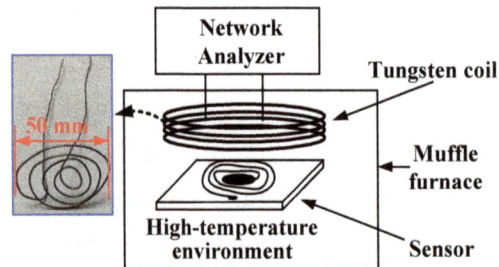

Figure 10. Measured frequency *versus* temperature.

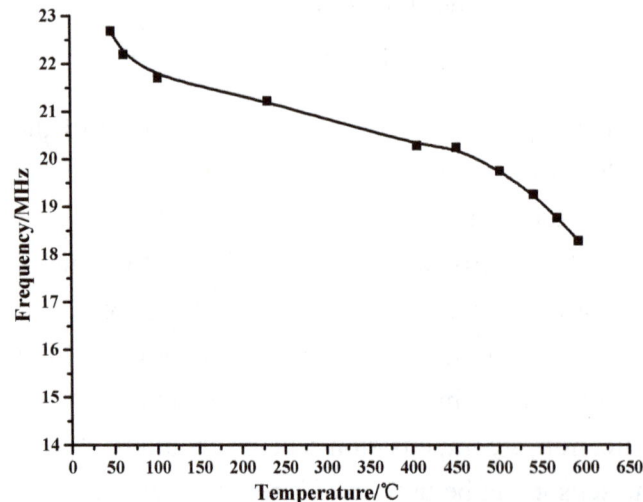

The quality factor of the sensor is defined as [14]:

$$Q = \frac{2\pi f_0 L_s}{R_s} = \frac{1}{R_s}\sqrt{\frac{L_s}{C_s}} \tag{2}$$

As illustrated in Figure 11, it is clear that the sensor realized by post-fire metallization with Ag paste has a better coupling effect compared to the sensor fabricated by cofiring with Pt paste. This is because Ag has a smaller resistivity than Pt. The resistivity of Ag at room temperature is 1.58×10^{-8} $\Omega \cdot m$ and the resistivity of Pt is 1.1×10^{-7} $\Omega \cdot m$. The resistivity of metal at a temperature T can be expressed as:

$$\rho_T = \rho_0(1 + \alpha T) \tag{3}$$

where ρ_0 is the resistivity at room temperature, and α is the temperature coefficient of the resistivity. The temperature coefficient for Ag and Pt are 0.0038 and 0.00374, respectively. The direct-current (DC) resistance of the Ag inductance coil at room temperature measured by a multimeter is 5.7 Ω, and corresponding resistance of the cofired sensor is 39.7 Ω. By using the Equation (3), the resistance of the sensor at 600 °C is 18.696 Ω, which is less than that of the sensor cofired with Pt paste. Therefore, the sensor fabricated by the proposed method comparatively has an excellent coupling effect at high temperature.

Figure 11. Measured impedance phase *versus* frequency at 600 °C.

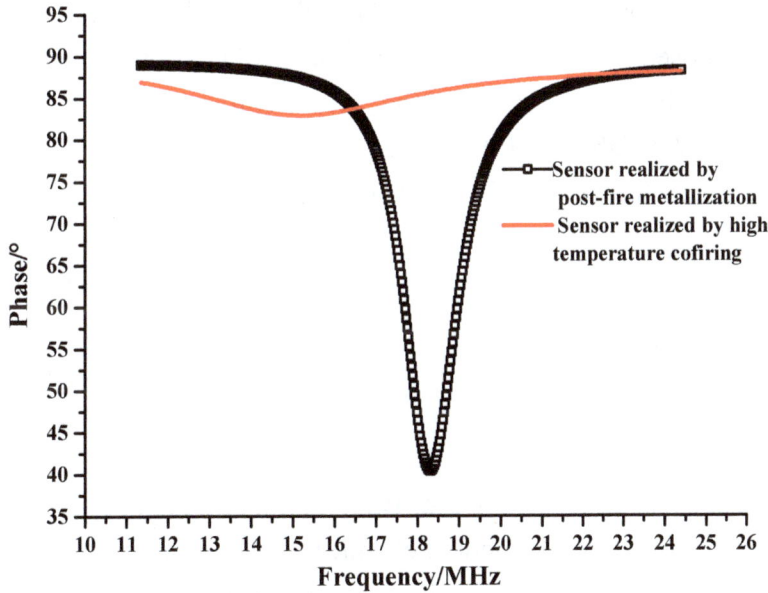

As shown in Figure 11, the phase dip corresponding to the proposed sensor at 600 °C is approximately 50°, which is quite considerable compared with the phase dip of 6° for the sensor fabricated by cofiring with Pt paste. The larger phase dip of the proposed sensor at 600 °C can extend its working temperature to a higher point. Although zirconia ceramic has mechanical stability at the temperature above 600 °C, the sensor realized by co-firing cannot be used at a higher temperature owing to the weak coupling. Therefore, the proposed method can make a zirconia-ceramic-based passive pressure sensor for application in higher temperature environments.

5. Conclusions

The detailed design, fabrication, and testing of a zirconia-ceramic-based passive pressure sensor fabricated by post-fire metallization was presented in this article. The sensor realized by the proposed method has an advantage over existing cofired sensors in the high-temperature operation range due to its comparable excellent coupling effect. Experimental results showed that the sensor resonance frequency exhibited a considerable average sensitivity of 790 kHz/bar within the measurement range of 0–1 bar. By comparing the corresponding phase dip of the proposed sensor and a sensor previously fabricated by cofiring with Pt paste at 600 °C, it is clear that the proposed sensor can function at temperatures greater than 600 °C; however, the sensor realized by cofiring cannot operate in higher temperature environments owing to its weak coupling.

Acknowledgments

This work was supported by the National Natural Science Foundation of China (Grant Nos. 61471324 and 51205373), the Program for the Outstanding Innovative Teams of Higher Learning Institutions of Shanxi, and the Shanxi Natural Science Foundation (Grant No. 2012021013-4). The authors would also like to thank the anonymous reviewers for their valuable comments and suggestions in improving the quality of this paper.

Author Contributions

All works with relation to this paper have been accomplished by all authors' efforts. The idea and design of the sensor were proposed by Tao Luo. The experiments of the sensor were completed with the help from Liqiong Ding and Chao Li. Chenyang Xue gave significant guidance on the selection of sintering curve, and Tanyong Wei fabricated the sensor. At last, every segment relate to this paper is accomplished under the guidance from Qiulin Tan and Jijun Xiong.

Conflicts of Interest

The authors declare no conflict of interest.

References

1. Hunter, G.W.; Neudeck, P.G.; Liu, C.C.; Ward, B.; Wu, Q.H.; Dutta, P.; Frank, M.; Trimbol, J.; Fulkerson, M.; Patton, B.; *et al.* Development of chemical sensor arrays for harsh environments and aerospace applications. In Proceedings of the IEEE Sensors, Orlando, FL, USA, 12–14 June 2002; Volume 2, pp. 1126–1133.

2. Van Netten, C. Design of a small personal air monitor and its application in aircraft. *Sci. Total Environ.* **2009**, *407*, 1206–1210.

3. Baptista, C.A.R.P.; Barboza, M.J.R.; Adib, A.M.L.; Andrade, M.; Otani, C.; Reis, D.A.P. High temperature cyclic pressurization of titanium ducts for use in aircraft pneumatic systems. *Mater. Des.* **2009**, *30*, 1503–1510.

4. Xu, J.; Pickrell, G.; Wang, X.; Peng, W.; Cooper, K.; Wang, A. A novel temperature-insensitive optical fiber pressure sensor for harsh environments. *IEEE Photonics Technol. Lett.* **2005**, *17*, 870–872.

5. Kroetz, G.H.; Eickhoff, M.H.; Moeller, H. Silicon compatible materials for harsh environment sensors. *Sens. Actuators A* **1999**, *74*, 182–189.

6. Senesky, D.G.; Jamshidi, B.; Cheng, K.B.; Pisano, A.P. Harsh environment silicon carbide sensors for health and performance monitoring of aerospace systems: A review. *IEEE Sens. J.* **2009**, *9*, 1472–1478.

7. Radosavljevic, G.J.; Zivanov, L.D.; Smetana, W.; Maric, A.M.; Unger, M.; Nad, L.F. A wireless embedded resonant pressure sensor fabricated in the standard LTCC technology. *IEEE Sens. J.* **2009**, *9*, 1956–1962.

8. Meijerink, M.G.H.; Nieuwkoop, E.; Veninga, E.P.; Meuwissen, M.H.H.; Tijdink, M.W.W.J. Capacitive pressure sensor in post-processing on LTCC substrates. *Sens. Actuators A* **2005**, *123–124*, 234–239.

9. Tan, Q.; Kang, H.; Xiong, J.; Qin, L.; Zhang, W.; Li, C.; Ding, L.; Zhang, X.; Yang, M. A wireless passive pressure microsensor fabricated in HTCC MEMS technology for harsh environments. *Sensors* **2013**, *13*, 9896–9908.

10. Andrei, A.; Malhaire, C.; Brida, S.; Barbier, D. Long-term stability of metal lines, polysilicon gauges, and ohmic contacts for harsh-environment pressure sensors. *IEEE Sens. J.* **2006**, *6*, 1596–1601.

11. Du, J.; Zorman, C.A. A polycrystalline SiC-on-Si architecture for capacitive pressure sensing applications beyond 400 °C: Process development and device performance. *J. Mater. Res.* **2013**, *28*, 120–128.

12. Marsi, N.; Majlis, B.Y.; Hamzah, A.A.; Mohd-Yasin, F. The mechanical and electrical effects of MEMS capacitive pressure sensor based 3C-SiC for extreme temperature. *J. Eng.* **2014**, *2014*, 715167.

13. Okojie, R.S.; Lukco, D.; Nguyen, V.; Savrun, E. High temperature SiC pressure sensors with low offset voltage shift. *SPIE Proc.* **2014**, *9113*, 911308.

14. Fonseca, M.A. Polymer/Ceramic Wireless MEMS Pressure Sensors for Harsh Environments: High Temperature and Biomedical Applications. Ph.D. Thesis, Georgia Institute of Technology, Atalanta, GA, USA, 2007.

2

A Particle Filter for Smartphone-Based Indoor Pedestrian Navigation

Andrea Masiero *, Alberto Guarnieri †, Francesco Pirotti † and Antonio Vettore †

CIRGEO (Interdepartmental Research Center of Geomatics), University of Padova,
via dell'Università 16, 35020 Legnaro (PD), Italy; E-Mails: alberto.guarnieri@unipd.it (A.G.);
francesco.pirotti@unipd.it (F.P.); antonio.vettore@unipd.it (A.V.)

† These authors contributed equally to this work.

* Author to whom correspondence should be addressed; E-Mail: masiero@dei.unipd.it

External Editors: Naser El-Sheimy and Aboelmagd Noureldin

Abstract: This paper considers the problem of indoor navigation by means of low-cost mobile devices. The required accuracy, the low reliability of low-cost sensor measurements and the typical unavailability of the GPS signal make indoor navigation a challenging problem. In this paper, a particle filtering approach is presented in order to obtain good navigation performance in an indoor environment: the proposed method is based on the integration of information provided by the inertial navigation system measurements, the radio signal strength of a standard wireless network and of the geometrical information of the building. In order to make the system as simple as possible from the user's point of view, sensors are assumed to be uncalibrated at the beginning of the navigation, and an auto-calibration procedure of the magnetic sensor is performed to improve the system performance: the proposed calibration procedure is performed during regular user's motion (no specific work is required). The navigation accuracy achievable with the proposed method and the results of the auto-calibration procedure are evaluated by means of a set of tests carried out in a university building.

Keywords: indoor navigation; positioning; sensor fusion; nonlinear filtering; smartphones geolocation

1. Introduction

Thanks to certain socially relevant applications (e.g., localizing and tracking people inside buildings during emergencies), indoor navigation is recently becoming a topic of wide interest among the research community. Despite the recent developments in localization with wireless sensor networks (e.g., by means of the radio signal strength (RSS), [1–7]) and in pedestrian tracking (e.g., urban dead-reckoning, [8,9]), indoor navigation is still considered an open problem [10–13]. Specifically, three challenging aspects can be recognized in indoor navigation: the required accuracy is typically higher than in outdoor applications, the GPS signal is typically not available and there is low reliability of the sensor measurements (e.g., inertial navigation system (INS) measurements and RSS).

The main issue in the use of INS measurements to provide updates of the estimated position is that such an estimate quickly drifts from the real one. On the other hand, position estimation based on RSS measurements does not suffer from drifts; however, because of environment variability and signal instability, the estimation error of such an approach is typically quite large.

Since the use of none of the considered approaches (INS and RSS) can allow by itself obtaining a sufficiently good estimation error in the considered conditions of interest, a commonly adopted strategy to tackle this problem relies on the integration between the data collected by several types of sensors to achieve more robust localization results [14–17]. In particular, in this paper, an indoor navigation system with minimal positioning sensor equipment is considered: the goal of this work is to enable navigation with low-cost mobile devices (typically carried by the user's hand) in indoor and other critical environments, e.g., the proposed navigation algorithm can be executed on a smartphone, which estimates its own position inside a building by combining the information collected from the Wi-Fi network (RSS) with measurements derived from the embedded INS sensor (the smartphone considered here is provided with a three-axis accelerometer and three-axis magnetometer). Furthermore, a map of the building is assumed to be available to the tracking algorithm: this information can be either provided by the Wi-Fi network or acquired by image plots or plans of the building.

Similarly to other indoor navigation methods recently proposed in the literature [17,18], in this work, the tracking problem is faced from a probabilistic point of view: the probability distribution of the current position is described as a sum of particles whose movements are determined by the sensor measurements (see [19–21] for a description of particle filters and [22–25] for some particle filtering applications).

The approach adopted in this paper is actually a generalization of the particle filter proposed in [17]: the proposed changes aim at compensating measurement errors due to the use of uncalibrated sensors. Indeed, in order to make the use of the system as simple as possible to the user, sensors are not calibrated by using dedicated specific procedures. Instead, the particle filter proposed in [17] is properly modified to deal with measurements from uncalibrated sensors. Then, an auto-calibration procedure is adopted for the magnetic sensor to improve the system performance. Differently from calibration algorithms typically used in similar systems [26–28], this calibration procedure does not require any specific action by the user: it is performed during the navigation, without affecting the user's regular moves in any way.

This way, the adopted navigation procedure results in a new algorithm that integrates the positional information derived from INS and Wi-Fi RSS measurements with the geometrical information provided by the building map, while dealing with uncalibrated sensors, as well.

The considered method has shown a significant improvement in the navigation accuracy with respect to the particle filter proposed in [17] on some experiments carried out in a university building using the standard Wi-Fi network.

The paper is organized as follows: Section 2 provides a basic description of the navigation system and its relation to [17] and other previous works. Section 3 introduces the tracking and sensor data fusion approach used in this work. Section 4 presents the procedure adopted to perform the auto-calibration of the magnetometer. Finally, in Section 5, the performance of the method is experimentally validated.

2. System Description and Relation to Previous Works

Thanks to the technological developments in electronics and MEMS-based sensors [29–32], nowadays, most mobile devices (e.g., smartphones and tablets) are provided with several embedded sensors. Hence, differently from other works, which considered the use of external sensors (e.g., [8,9,17]), here, an indoor navigation system based on the use of the sensors embedded in standard mobile devices is considered.

To be more specific, a minimal equipment of embedded sensors is required: a three-axis accelerometer, a three-axis magnetometer (e.g., MEMS-based inertial navigation system) and a Wi-Fi antenna. Furthermore, the navigation system is assumed to have access to the standard Wi-Fi network available inside the building. The installation of additional access points (APs) is not required (access points on one of floors of the considered building are shown as red circles in Figure 1a).

Figure 1. (a) Coordinate system in the university building used as a test field, and (b) the smartphone coordinate system (u_s, v_s, w_s). Smartphone used for the test: Huawei Sonic U8650.

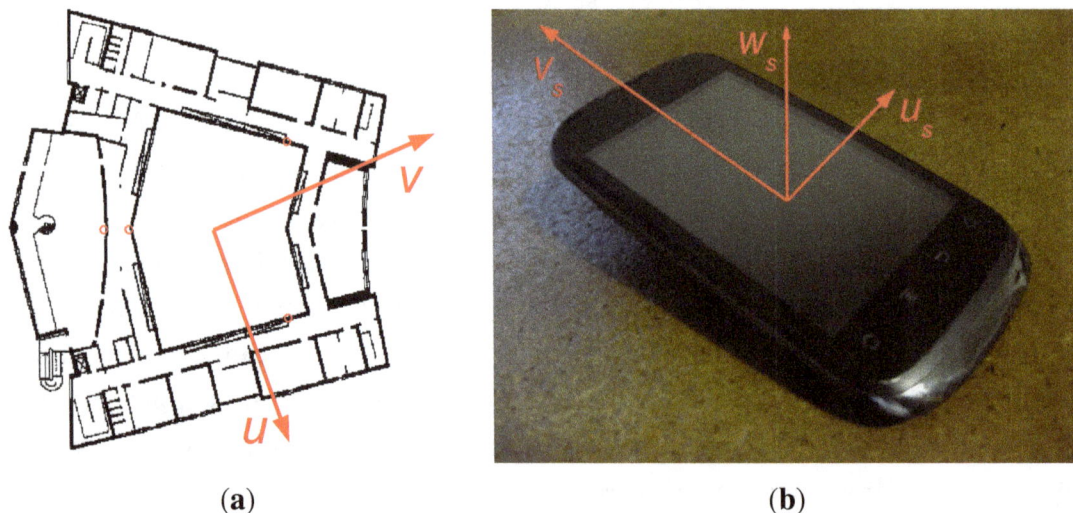

(a)   (b)

In addition, to make the use of the system as simple as possible from the user's point of view, no specific initial sensor calibration procedure is needed: the tracking algorithm that will be presented in Section 3 is designed to deal with uncalibrated sensors. Furthermore, as shown in Section 4, in order to improve the navigation performance, the system is designed to auto-calibrate the magnetic sensor during the navigation, e.g., exploiting regular user's movements.

The rationale is that of using a dead reckoning-like approach [8,9]: detect the human steps by means of a proper analysis of the accelerometer measurements [33], then use the combination of magnetometer and accelerometer measurements to estimate the movement direction with respect to the north [28].

Since, in terrestrial applications, usually, the height with respect to the floor is of minor interest, then, hereafter, navigation is considered as a planar tracking problem (exceptions to this assumption are admitted, e.g., to deal with stairs and lifts). Let (u_t, v_t) be the position of the device (e.g., smartphone), expressed with respect to the north and east directions (*i.e.*, global reference system; Figure 1a shows the coordinate system in a university building used as the test field), before the t-th step; then:

$$\left[\begin{array}{c} u_{t+1} \\ v_{t+1} \end{array}\right] = \left[\begin{array}{c} u_t \\ v_t \end{array}\right] + s_t \left[\begin{array}{c} \sin \alpha_t \\ \cos \alpha_t \end{array}\right] \qquad (1)$$

where s_t is the length of the t-th step and α_t is the corresponding heading direction. An estimation of the initial position (u_0, v_0) is assumed to be (*a priori*) available.

The step length s_t is assumed to be provided by a proper analysis of the accelerator measurements [33]. Alternatively, s_t can be fixed to a constant value (an approximation of the mean step length): the tracking algorithm described in Section 3 is designed to compensate (relatively small) step length errors.

In most of the cases of common interest, the mobile device is supposed to be carried by the user's hand (this is the case considered in the simulations of Section 5); however, the navigation system is expected to work in different conditions, as well. To be more specific, the system is expected to work as long as the heading direction is approximately fixed with respect to the local coordinate system (more details on this assumption are provided in the Appendix) (u_s, v_s, w_s) (the local coordinate system for a smartphone used in the simulations of Section 5 is shown in Figure 1b). However, small deviations are allowed: the system of Section 3 is designed to estimate and correct heading direction discrepancies, with an absolute value lower than 36 degrees, with respect to the conventional direction.

Without loss of generalization, hereafter, the v_s and the w_s axes will be assumed to approximately correspond with the heading direction and with the vertical direction, respectively. Notice that these assumptions are not requirements of the proposed approach, but will ease the presentation of the method and the analysis of the calibration results shown in Section 5.

Notice that the update of Equation (1) according to the sensor measurements provides an approximation of the device trajectory. However, it is well known that, because of measurement errors, the trajectory estimated by means of Equation (1) quickly drifts from the correct one. In order to make the estimation method more robust, information provided by the dead reckoning-like approach has to be integrated with that provided by RSS measurements and the geometrical characteristics of the building (Section 3). RSS measurements of the standard Wi-Fi networks are provided by the corresponding antenna in the smartphone. The geometrical characteristics of the building are assumed to be pre-loaded on the navigation device or downloaded by means of the Wi-Fi network at the beginning of the navigation. As the mobile device gets inside of the building, it starts communicating with the Wi-Fi network, hence the device is considered as immediately ready to go (the total time for connecting the device to the network and to receive the information required by the navigation system depends on the specific conditions of the Wi-Fi network (e.g., signal strength, connection data rate, number of users connected to the network); however, the communication step is typically very fast (e.g., quantity of information to be passed < 100 kBytes in our simulations)).

Several nonlinear filtering strategies have been previously considered in the literature for the integration of information (e.g., INS and RSS measurements) in pedestrian indoor navigation [8,9,16,17]. Among such possible strategies, a particle filter approach [19] is considered. In particular, our approach can be considered as a generalization of the particle filter proposed in [17,34], and that will be briefly summarized in the following (the reader is referred to [17,34] for a detailed description).

2.1. Widyawan et al. Particle Filter

Particle filter techniques are statistical methods widely used in complex (possibly nonlinear) estimation problems [19–24]. The variable of interest at time t, q_t (e.g., the device position in the navigation problem considered here), is described by means of a density function $p(q_t|Y_t)$, which depends on the values of the past measurements Y_t: the goal of the particle filter is to properly estimate such posterior density $p(q_t|Y_t)$ (and update it when new measurements are available). Particle filters are particularly effective when dealing with complex systems, because of the two following characteristics:

- The probability function $p(q_t|Y_t)$ is not restricted to be Gaussian nor univariate; instead, it is described as the sum of a set of n particles.
- No restriction is imposed on the dynamics of the system (*i.e.*, it can deal with both linear and nonlinear systems).

However, when dealing with complex systems (e.g., nonlinear and with non-univariate density $p(q_t|Y_t)$), a large number of particles n has to be used in order to ensure good filtering performance: hence, the ability to deal with difficult scenarios is paid for with an increase of the computational complexity of the filter.

Several particle filter methods have been proposed in the literature, and each of them can be summarized as a specific algorithm used in order to properly updated the particle description of $p(q_t|Y_t)$ when new measurements are available. The rest of this section is dedicated to the description of the particle filter for the indoor navigation proposed in [17,34], while a generalization of such an approach will be presented in Section 3 in order to efficiently deal with uncalibrated sensors (while keeping the number of particles n relatively small).

Let y_t be the vector of measurements corresponding to the t-th step and Y_t be the collection of measurements $\{y_\tau\}$ from $\tau = 0$ to $t - 1$. Furthermore, let $q_t = [u_t \ v_t]^\top$; then the probability distribution of estimated position q_t after the $(t - 1)$-th step is expressed as follows:

$$p(q_t|Y_t) = \sum_{i=1}^{n} w_{i,t}\delta(q_t - q_{i,t}) \tag{2}$$

where $q_{i,t}$ and $w_{i,t}$ are the position and the weight of the i-th particle at t, while $\delta(\cdot)$ is the Dirac delta function.

In order to simplify the presentation, with a slight abuse of notation, hereafter, the subscript index t will be referred to as a time index, as well.

In this subsection, the INS sensor measurements are assumed to be unbiased, but affected by zero-mean random Gaussian noise, with heading direction and step length variance σ_α^2 and σ_s^2, respectively. Then, the position of the i-th particle at time $t+1$ is updated similarly to Equation (1):

$$\hat{q}_{i,t+1} = q_{i,t} + s_{i,t} \begin{bmatrix} \sin \alpha_{i,t} \\ \cos \alpha_{i,t} \end{bmatrix} \tag{3}$$

where $\alpha_{i,t}$ and $s_{i,t}$ are randomly sampled from the proper Gaussian distributions: $\alpha_{i,t} \sim \mathcal{N}(\alpha_t, \sigma_\alpha^2)$, $s_{i,t} \sim \mathcal{N}(s_t, \sigma_s^2)$. σ_α and σ_s are set to proper constant values ($\sigma_\alpha = \pi/2$, $\sigma_s = 0.15$ m in [17]).

Exploiting an RSS channel model [3], the RSS measurements are converted to distance measurements. Typically, the error on each measured distance is assumed to be normally distributed [3,17]. Then, the comparison of measured distances with the distances from the access points to the estimated device position yields the RSS probability $p_{RSS}(\hat{q}_{i,t+1}|y_t)$, that is the probability of being positioned in $\hat{q}_{i,t+1}$ given the RSS measurement in y_t (see [3,17] for details on $p_{RSS}(\cdot)$).

Then, the particle weights are updated as follows:

$$\hat{w}_{i,t+1} = \begin{cases} p_{RSS}(\hat{q}_{i,t+1}|y_t) & \text{if } y_t \text{ contains an RSS measurement} \\ 1/n & \text{otherwise,} \end{cases} \tag{4}$$

Furthermore, particles with trajectories that violate the geometric constraints of the building are discarded. Then, the weights $\{\hat{w}_{i,t+1}\}_{i=1,\ldots,n}$ are scaled in order to be normalized to one, $i.e.$, $\sum_{i=1,\ldots,n} \hat{w}_{i,t+1} = 1$, and the following distribution for the particles at time $t+1$ is obtained:

$$p(q_{t+1}|Y_{t+1}) \approx \sum_{i=1}^{n} \hat{w}_{i,t+1}\delta(q_{t+1} - \hat{q}_{i,t+1}) \tag{5}$$

In the above equation, part of the particles may have very low weights (zero if the particle trajectory violates the building constraints); then, the effective number $\hat{n}$ of particles in Equation (5) is typically lower than n (a detailed study of the effective number of particles goes beyond the goal of this paper; for simplicity of exposition, hereafter, $\hat{n}$ at time t will be considered as the number of particles, such that $w_{i,t}$ is significantly greater than zero ($i.e.$, larger than a given threshold); the reader is referred to [20,21] for a more detailed description of particle filter characteristics). In order to preserve the effectiveness of the particle representation, the final approximation of the position distribution at time $t+1$ is obtained by resampling n particles from Equation (5), $i.e.$, the following actions has to be done for each resampled particle i': an index i is randomly sampled accordingly to the probability $\hat{w}_{i,t+1}$; then, $q_{i',t+1}$ is sampled from a Gaussian distribution centered at $\hat{q}_{i,t+1}$, and its weight is set to $w_{i',t+1} = 1/n$ (notice that the resampling step increases the estimation variance; see [35] for a comparison of different resampling schemes).

The above approach can be summarized with the following procedure to be executed each time a new measurement is available:

(1) For each particle i:

    (1.a) draw a sample $\hat{q}_{i,t+1}$ from the proposal distribution ($i.e.$, draw samples $\alpha_{i,t}$ and $s_{i,t}$ from $\mathcal{N}(\alpha_t, \sigma_\alpha^2)$ and $\mathcal{N}(s_t, \sigma_s^2)$, respectively, then compute $\hat{q}_{i,t+1}$ with Equation (3)).

(1.b) Set $\hat{w}_{i,t+1} = 0$ if $\hat{q}_{i,t+1}$ violates building geometrical constraints; otherwise, set the weight $\hat{w}_{i,t+1}$ as in Equation (4).

(2) Scale the particle weights in order to normalize their sum to one: $\sum_{i=1,\dots,n} \hat{w}_{i,t+1} = 1$.

(3) Resample n particles from Equation (5) and set $w_{i',t+1} = 1/n$, $\forall i'$.

Notice that the number of operations to be executed for each particle in the above procedure can be upper bounded by a proper constant (and lower bounded, as well); hence, the computational complexity of the above particle filter is linear with respect to the number of particles n.

The reader is referred to [17,34] for details on the implementation of this particle filtering approach.

2.2. Issues Related with the Use of Uncalibrated Sensors

The particle filter summarized in the previous subsection has been proposed in [17,34] for devices with calibrated sensors. In such conditions, measurement errors can be assumed to be mostly random: in this case, the random sampling of the heading direction, $\alpha_{i,t} \sim \mathcal{N}(\alpha_t, \sigma_\alpha^2)$, and of the step length, $s_{i,t} \sim \mathcal{N}(s_t, \sigma_s^2)$, and the filtering obtained by means of imposing the geometrical constraints of the building (Step 1.b) of the procedure presented in the previous subsection) typically allow one to properly deal with these errors.

However, when working with uncalibrated sensors, measurement errors are typically correlated and larger (in absolute value) with respect to the calibrated case. This makes much more challenging the particle filter tracking task: this often leads to a significant decrease of $\hat{n}$ and, if n is not sufficiently large, to a tracking fault of the particle filter, *i.e.*, none of the particles describe adequately the device position; hence, the particle filter loses the correct track (this case occurred quite often in our simulations; in order to reduce the occurrence of such case, then, in our simulations of the particle filter described in Subsection 2.1, Step 1.a has been repeated (for a maximum of five times) when the generated sample $\hat{q}_{i,t+1}$ violates building geometrical constraints at Step 1.b).

It is well known that using a sufficiently large number of particles n, particle filtering techniques allow one to approximate arbitrarily well the considered system. However, the considered tracking algorithm is supposed to run in real time on a mobile device with limited resources (*i.e.*, computational power and available energy); hence, the number of particles n should be as small as possible in order to reduce the computational load and the power consumption of the device.

The above considerations motivate the generalization that will be presented in the following section of the particle filter of Subsection 2.1 in order to improve its performance (e.g., allow one to obtain similar results with a smaller number of particles) in the uncalibrated sensors case.

3. Tracking with Uncalibrated Sensors

The use of uncalibrated sensors can introduce temporally correlated measurement errors for both the heading direction and the step length. The following two subsections will separately consider these two cases, while Subsection 3.3 will summarize the overall approach.

3.1. Heading Direction Correction

The heading direction measurement α_t can be statistically described as follows:

$$\alpha_t = \bar{\alpha}_t + \alpha_{bs,t} + \alpha_{be,t} + e_{\alpha,t} \tag{6}$$

where $\bar{\alpha}_t$ is the correct heading direction, $\alpha_{bs,t}$ is an angular bias due to the use of an uncalibrated magnetometer, $\alpha_{be,t}$ is an angular bias due to local perturbations of the terrestrial magnetic field typically due to the presence of ferromagnetic materials in the environment and $e_{\alpha,t}$ is a temporally random measurement error (that can typically be assumed to be white and Gaussian). The goal of this subsection is to approximately compensate for the bias $\alpha_{b,t} = \alpha_{bs,t} + \alpha_{be,t}$.

The following considerations are now in order:

- When the device moves along a rectilinear trajectory, *i.e.*, $\bar{\alpha}_t$ is (approximately) fixed to a constant value for $t = t_s, \ldots, t_e$, then $\alpha_{bs,t} \approx \alpha_{bs,t'}$ for $t_s \le t \le t_e, t_s \le t' \le t_e$.
- Assuming the magnetic field to be sufficiently smooth in the considered environment, then locally $\alpha_{be,t} \approx \alpha_{be,t'}$ for $t \approx t'$.
- Human trajectories are usually quite regular.

Then, the rationale is that of decomposing the device motion in piecewise (approximately) rectilinear trajectories. Each piece of such trajectories being approximately rectilinear and typically not so long, then:

$$(\alpha_{bs,t} + \alpha_{be,t}) \approx (\alpha_{bs,t'} + \alpha_{be,t'}) \tag{7}$$

for t and t' in the same piece of (approximately) rectilinear trajectory.

The end of a piece of the (approximately rectilinear) trajectory (and consequently, the beginning of a new piece) is detected when the following condition holds:

$$|\alpha_t - \alpha_{t-1}| > \alpha_{thr} \ \vee \ |\alpha_t - \alpha_{t-2}| > \alpha_{thr} \ \vee \ |\alpha_t - \alpha_{t-3}| > \alpha_{thr} \tag{8}$$

where $\vee$ indicates the logic operator "or" and α_{thr} is a proper threshold ($\alpha_{thr} = \pi/6$ in the simulations of Section 5).

Accordingly to Condition (8), the trajectory is divided in non-overlapping sections of (approximately) rectilinear trajectories. Let $t_{s,j}$ and $t_{e,j}$ be the initial and ending time instant for the j-th (approximately rectilinear) section. Then, in order to compensate for the bias $\alpha_{b,t}$, $t_{s,j} \le t \le t_{e,j}$, Step 1.b of the particle filtering procedure presented in Subsection 2.1 is modified as follows:

(1.b.1) If $\hat{q}_{i,t+1}$ violates the geometrical constraints of the building, then resample $\hat{q}_{i,t+1}$ as follows: draw a rotation angle sample $\hat{\alpha}_{b,t}$ from $\mathcal{U}(-\alpha_{\max}, \alpha_{\max})$, and rotate $(\hat{q}_{i,t_{s,j}}, \ldots, \hat{q}_{i,t_{e,j}})$ of an angle $\hat{\alpha}_{b,t}$ around the vertical axis passing through $\hat{q}_{i,t_{e,j-1}}$.

(1.b.2) Set $\hat{w}_{i,t+1} = 0$ if $\hat{q}_{i,t+1}$ violates building geometrical constraints; otherwise, set the weight $\hat{w}_{i,t+1}$ as in Equation (4).

Since the angular bias $\alpha_{b,t}$ is approximately constant over an (approximately rectilinear) section of the trajectory $t_{s,j} \le t \le t_{e,j}$, then Step 1.b.1 aims at compensating for such angular bias (if present), subtracting a randomly sampled angular bias $\hat{\alpha}_{b,t}$ to the heading direction measurements over all of the trajectory section $t_{s,j} \leq t \leq t_{e,j}$. Since no information about such an angular bias is available, $\hat{\alpha}_{b,t}$ is randomly sampled from a uniform distribution in the interval $(-\alpha_{\max}, \alpha_{\max})$, i.e., $\mathcal{U}(-\alpha_{\max}, \alpha_{\max})$, where $\alpha_{\max} = \pi/5$ in our simulations of Section 5.

Then, Step 1.b.2, which corresponds to Step 1.b in Subsection 2.1, has the task of properly filtering out particles associated with improper values of $\hat{\alpha}_{b,t}$ (and of the other trajectory parameters).

3.2. Step Length Correction

The use of an uncalibrated accelerometer leads to a possible bias in the step length measurements: $s_t = \bar{s}_t + s_{b,t} + e_{s,t}$, where s_t is the measured step length, $\bar{s}_t$ is the correct value of the step length, $s_{b,t}$ is the bias and $e_{s,t}$ is a zero-mean white Gaussian noise with variance σ_s^2.

In order to maximize the flexibility of the filter, but without excessively affecting the complexity of the filtering procedure, analogous to the previous Subsection , it is assumed that the value of the step length bias can be different in different sections of the trajectory; however, it is assumed (approximately) constant within each section (for simplicity of the filtering procedure, the same trajectory sections determined in Subsection 3.1 are considered here; nevertheless, the determination of a possible relation between the characteristics of human walking and of the step length bias will be the subject of a future investigation), i.e., $s_{b,t} \approx s_{b,t'}$ for $t_{s,j} \leq t \leq t_{e,j}, t_{s,j} \leq t' \leq t_{e,j}$.

Then, Step 1.a of Subsection 2.1 is modified as follows in order to take into account a possible step length bias:

(1.a) Draw a sample $\hat{q}_{i,t+1}$ from the proposal distribution: draw samples $\alpha_{i,t}$ and $s_{i,t}$ from $\mathcal{N}(\alpha_t, \sigma_\alpha^2)$ and $\mathcal{N}(s_t + \hat{s}_{b,i,t}, \sigma_s^2)$, respectively; then, compute $\hat{q}_{i,t+1}$ with Equation (3)).

where, for each particle i, $\hat{s}_{b,i,t}$ is randomly drawn from a uniform distribution $\mathcal{U}(-s_{\max}, s_{\max})$ at the beginning of each section of the trajectory. $\hat{s}_{b,i,t}$ is fixed to such a value for all of the time instants of the same section of the trajectory, i.e., $\hat{s}_{b,i,t} = \hat{s}_{b,i,t'}$ for $t_{s,j} \leq t \leq t_{e,j}, t_{s,j} \leq t' \leq t_{e,j}$.

Then, as previously shown, Step 1.b.2 has the task of properly filtering out particles associated with improper values of $\hat{s}_{b,i,t}$ (and of the other trajectory parameters).

Finally, it is worth noticing that, when only INS measurements are available, the particle filter of Subsection 2.1 in open space results in being a biased estimator. For simplicity of exposition, assume that both heading direction and step length measurements are unbiased (or have already been properly compensated for). Furthermore, without loss of generality, it is assumed that $\bar{\alpha}_t = 0$ and that the device moves at time t from the origin of the global coordinate system, i.e., $q_t = [0\ 0]^\top$.

Then, the following condition holds for an unbiased estimator:

$$\mathcal{E}[q_{t+1}|Y_{t+1}] = \begin{bmatrix} 0 \\ \bar{s}_t \end{bmatrix} \tag{9}$$

where $\mathcal{E}[\cdot]$ indicates the expectation operator.

Since the α_t distribution is symmetric with respect to zero, it is simple to show that $\mathcal{E}[q_{t+1}|Y_{t+1}] = \begin{bmatrix} 0 \\ s \end{bmatrix}$ for a certain value of s. In order to compute s, consider the Taylor second order approximation of $\mathcal{E}[s]$ for $\alpha_t = e_{\alpha,t} \approx 0$ and $s_t = \bar{s}_t + e_{s,t} \approx \bar{s}_t$. Then,

$$\mathcal{E}[s_t \cos(\alpha_t)] \approx \bar{s}_t + \begin{bmatrix} \cos(0) & -\bar{s}_t \sin(0) \end{bmatrix} \begin{bmatrix} \mathcal{E}[e_{s,t}] \\ \mathcal{E}[e_{\alpha,t}] \end{bmatrix} + \frac{1}{2} \begin{bmatrix} -\sin(0) & -\bar{s}_t \cos(0) \end{bmatrix} \begin{bmatrix} \mathcal{E}[e_{s,t}^2] \\ \mathcal{E}[e_{\alpha,t}^2] \end{bmatrix} \quad (10)$$

$$\approx \bar{s}_t + 0 + \frac{1}{2} \begin{bmatrix} 0 & -\bar{s}_t \end{bmatrix} \begin{bmatrix} \sigma_s^2 \\ \sigma_\alpha^2 \end{bmatrix} = \bar{s}_t \left(1 - \frac{\sigma_\alpha^2}{2} \right) \quad (11)$$

Thus, for $\sigma_\alpha^2 > 0$, then $\mathcal{E}[s_t \cos(\alpha_t)] \neq \bar{s}_t$, and as previously claimed, the estimator is biased in open space.

However, in indoor navigation, it is also necessary to take into account the effect of the geometrical constraints of the building imposed in Step 1.b.2: the net effect of Step 1.b.2 is that of selecting those particles whose parameters are admissible with respect to the geometrical constraints of the building. Hence Step 1.b.2 leads the estimator towards the correction of its bias; however, this typically causes a decrease of the effective number of particles $\hat{n}$, which is a working condition to be preferably avoided, if possible.

Motivated by the above considerations, the step length measurements can be scaled by a constant factor k in order to compensate for the estimator bias:

$$k = \left(1 - \frac{\sigma_\alpha^2}{2} \right)^{-1} \quad (12)$$

Since, as mentioned above, the geometrical constraints of the building have a bias correction effect, from a practical point of view, it is possible to set k to different values related to Equation (12): for instance, in Section 5, k is set to $\frac{1}{2} \left(1 - \frac{\sigma_\alpha^2}{2} \right)^{-1}$.

3.3. Revised Particle Filter

The changes presented in the previous subsections with respect to the particle filter of Subsection 2.1 provide a revised particle filtering procedure that allows one to deal with uncalibrated sensors.

The overall particle filtering algorithm can be summarized as follows:

(1) For each particle i:

    (1.a') Draw a sample $\hat{q}_{i,t+1}$ from the proposal distribution: draw samples $\alpha_{i,t}$ and $s_{i,t}$ from $\mathcal{N}(\alpha_t, \sigma_\alpha^2)$ and $\mathcal{N}(ks_t + \hat{s}_{b,i,t}, \sigma_s^2)$, respectively; then, compute $\hat{q}_{i,t+1}$ with Equation (3).

    (1.b.1) If $\hat{q}_{i,t+1}$ violates the geometrical constraints of the building, then resample $\hat{q}_{i,t+1}$ as follows: draw a rotation angle sample $\hat{\alpha}_{b,t}$ from $\mathcal{U}(-\alpha_{\max}, \alpha_{\max})$, and rotate $(\hat{q}_{i,t_{s,j}}, \ldots, \hat{q}_{i,t_{e,j}})$ of an angle $\hat{\alpha}_{b,t}$ around the vertical axis passing through $\hat{q}_{i,t_{e,j-1}}$.

    (1.b.2) Set $\hat{w}_{i,t+1} = 0$ if $\hat{q}_{i,t+1}$ violates the building's geometrical constraints; otherwise, set the weight $\hat{w}_{i,t+1}$ as in Equation (4).

(2) Scale the particle weights in order to normalize their sum to one: $\sum_{i=1,\ldots,n} \hat{w}_{i,t+1} = 1$.

(3) Resample n particles from Equation (5) and set $w_{i',t+1} = 1/n$.

Notice that, similarly to the original filter of Subsection 2.1, the computational complexity of the revised particle filter is linear with respect to the number of particles n. However, its computational burden is increased with respect to the original version (by 10%–20%, approximately, in the simulations of Section 5), mostly because of Step 1.b.1.

4. Auto-Calibration of the Magnetic Sensor

The algorithm presented in the previous section allows one to obtain good tracking results also in the presence of uncalibrated sensors (as shown in the simulations of Section 5). Nevertheless, the use of calibrated sensors, if possible, is obviously preferred. This motivates the presentation in this section of an auto-calibration procedure for the magnetic sensor.

The rationale is that of collecting magnetic sensor measurements during the regular use of the device (*i.e.*, while the particle filter of Section 3 is running): compute the calibration of the magnetic sensor as soon as possible, and introduce the calibration model in the particle filter in order to improve its tracking ability.

Since the navigation problem has been considered as a 2D tracking problem and the heading direction has been assumed to be (approximately) fixed with respect to the local coordinate system, then the sensor calibration can be considered as a 2D calibration problem. To be more specific, since the heading direction and the vertical direction have been assumed to (approximately) correspond to the v_s and w_s axis, respectively, then the 2D calibration is performed on the (u_s, v_s) plane, which (approximately) corresponds to the horizontal plane.

Despite the above assumption, it is worth noticing that in practical applications, the attitude of the device is typically not fixed to a constant value for all of the time instants: in practice, the attitude of the device will be assumed to be only approximately fixed, e.g., it oscillates around a mean orientation.

On the one hand, from the above considerations, it is clear that since the (u_s, v_s) plane does not exactly correspond to the horizontal plane for all of the time instants, then even an extremely precise calibration of the magnetic sensor on such a plane cannot exactly compensate for the magnetometer error in all of the time instants. On the other hand, a complete (3D) calibration of the magnetic sensor requires a full set of rotations of the device, which actually should be avoided in order to simplify the use of the navigation system to the user. Thus, the aim of this section is just the computation of a rough calibration that allows one to reduce the magnetometer error (even if typically not completely compensating for it). The usefulness of this rough calibration will be shown in the results of Section 5.

The calibration procedure considered here is similar to that in [26,36,37].

The following model for the magnetic sensor measurements is considered:

$$y_{m,t} = Dm_t + d + e_{m,t} \tag{13}$$

where $y_{m,t}$ is the magnetic sensor measurement vector (projected on the (u_s, v_s) plane), m_t is the magnetic field orientation vector (projected on the horizontal plane and normalized to one), $e_{m,t}$ is the measurement error, while D and d contain the calibration parameters.

The goal of this section is that of estimating D and d from the measurements $\{y_{m,t}\}_{t=1,\ldots,T}$, by minimizing the sum of the squared norm of the errors:

$$(\hat{D}, \hat{d}, m_1, \ldots, m_T) = \arg\min \sum_{t=1}^{T} ||y_{m,t} - Dm_t - d||^2 \tag{14}$$

The solution of the above optimization problem will be computed by means of an iterative procedure that will be presented in the following.

First, in order to initialize the iterative optimization procedure, an initial estimation of the parameter values is required.

Since m_t is a vector with norm one, then an initial estimation of the calibration parameters can be computed similarly to [26]:

$$0 = ||m_t||^2 - 1 \approx ||D^{-1}(y_{m,t} - d)||^2 - 1 = y_{m,t}^{\top} A y_{m,t} + y_{m,t}^{\top} b + c \tag{15}$$

where $A = D^{-\top} D^{-1}$, $b = -2Ad$, $c = d^{\top} A d - 1$.

The equation $y_{m,t}^{\top} A y_{m,t} + y_{m,t}^{\top} b + c \approx 0$ can be rewritten as a linear equation: $Y\theta \approx 0$, where $\theta = [vec(A)^{\top} \ b^{\top} \ c]^{\top}$ (where $vec(A)$ indicates the vectorized version of A (notice that A is a symmetric matrix; hence, only a set of independent parameters of A should be included in θ)), while Y is a matrix properly formed by the magnetic sensor measurements (at different time instants).

The equation $Y\theta \approx 0$ can be solved (in least squares sense) by means of the singular value decomposition (SVD) [38], yielding a solution vector $\hat{\theta}$ containing a scaled version $\hat{A}$, $\hat{b}$, $\hat{c}$ of the parameters in A, b, c. The proper scale value h can be computed as in [26]: $h = (\hat{b}^{\top} \hat{A}^{-1} \hat{b} - \hat{c})^{-1}$. Then, with a slight abuse of notation, $\hat{A}$, $\hat{b}$, $\hat{c}$ are redefined as follows: $\hat{A} = h\hat{A}$, $\hat{b} = h\hat{b}$, $\hat{c} = h\hat{c}$.

The initial estimates $\hat{D}$ and $\hat{d}$ of the calibration parameters can be obtained as follows:

$$\hat{A}^{-1} = \hat{D}\hat{D}^{\top} \tag{16}$$

$$\hat{d} = -\frac{1}{2}\hat{A}^{-1}\hat{b} \tag{17}$$

where the factorization of the $\hat{A}^{-1}$ matrix (which can be computed for instance by means of the Cholesky factorization [38] or with the SVD) is not unique. Indeed, let $\hat{D}$ be the matrix resulting from the Cholesky factorization of $\hat{A}^{-1}$; then, the factorization $\hat{A}^{-1} = (\hat{D}R)(R^{\top}\hat{D}^{\top})$ holds, as well (for each rotation matrix R).

Notice that, analogously to [36], an estimate $\hat{R}$ of the rotation matrix R can be obtained by solving an orthogonal Procrustes problem [39]. However, such an estimate of $\hat{R}$ will not be necessary in the following.

Once initial estimates of the involved parameters are provided as above, then the following two steps are iteratively executed (until convergence) in order to compute the solution of Equation (14):

(1) Consider the values of $\hat{D}$ and $\hat{d}$ as fixed, and compute $\hat{m}_t = \hat{D}^{-1}(y_{m,t} - \hat{d})$, for $t = 1, \ldots, T$. Then, redefine $\hat{m}_t$, for each t, by normalizing its norm to one: $\hat{m}_t = \frac{\hat{m}_t}{||\hat{m}_t||}$.

(2) Consider the values of $\{\hat{m}_t\}_{t=1,\ldots,T}$ as fixed and compute $\hat{D}$ and $\hat{d}$ by minimizing:

$$\sum_{t=1}^{T} ||\hat{D}\hat{m}_t - y_{m,t} + \hat{d}||^2 \tag{18}$$

Since, in this step, both $\{\hat{m}_t\}_{t=1,\dots,T}$ and $\{y_{m,t}\}_{t=1,\dots,T}$ are considered as known, then the (least squares) solution of the above problem can be easily computed in matrix form, for instance, by means of the Moore–Penrose pseudo-inverse [38].

Notice that, as usual, when dealing with iterative optimization techniques, the convergence of the proposed two-step procedure to the global minimum of Equation (14) is not guaranteed.

The sensor model resulting from the magnetometer calibration procedure can be introduced in the particle filter presented in Section 3: when the calibration results are available, calibrated measurements are substituted to the $\{\alpha_t\}$, and Step 1.b.1 of the revised particle filter can be discarded (or can be executed, but setting $\alpha_{\max}$ to a smaller value).

Since the use of a calibrated magnetometer can improve the performance of the particle filter, the calibration model should be used as soon as possible, *i.e.*, as soon as it provides reliable results. The results of the calibration procedure presented above can be considered (roughly) reliable, provided that a minimal number of measurements (e.g., 20 in our simulations) are in all of the octants of the (u_s, v_s) plane: once such conditions are achieved, the resulting calibration model can be used in the tracking procedure. Then, the calibration procedure shall be repeated later during the tracking of the device in order to reduce the calibration errors due to the possible small number of measurements in certain octants.

5. Experimental Validation and Discussion

The experimental validation of the method proposed in the previous sections has been assessed using a low-cost smartphone, Huawei Sonic U8650, in a building at the University of Padova. The considered smartphone is provided with a three-axis accelerometer (Bosh, BMA150, resolution 0.15 m/s², maximum range ±39.24 m/s²), a three-axis magnetic field sensor (Asahi Kasei, AK8973, resolution 0.0625 μT, maximum range ±2000 μT), and a Wi-Fi receiver 802.11 b/g/n. In the application used for the simulations presented in this Section, sensor measurements are acquired when changed values are detected by the system (e.g., by means of the Android sensor event method *onSensorChanged*); consequently, sensor data rates are not constant. In the simulations presented here, the approximate values of the average data rates are: 5 Hz for the accelerometer, 7 Hz for the magnetometer (since several magnetometer measurements are typically available for each human step, the heading direction is computed as the average angle obtained from such measurements) and 0.2 Hz for the Wi-Fi network (notice that such values are just indicative, because real sensor data rate values can be influenced by many factors).

A set of 30 check points has been distributed along a track involving two floors of the building: Figure 2 shows part of the trajectory (green solid line) and the check points positions (red circles) along the track. Other different tracks have been considered, as well: despite the numerical results being obviously slightly different, the performance trend is analogous to that presented here. In this sense, the case shown here can be considered as a general case (representative of the typical performance of the method).

Figure 2. Path traveled by the volunteers on the first floor of the building (green solid line) and check point positions (red circles) along the track.

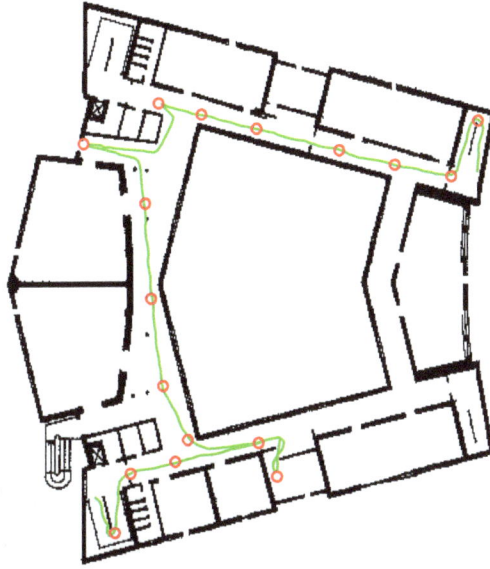

In order to make the results independent of a specific stride, experimental data have been collected by three volunteers, two men and one woman, with heights from 1.65 m to 1.85 m, which completed the track with a number of steps ranging from 300 to 350. Actually, the results obtained by using the data collected by the three volunteers are quite similar: since no peculiarity has been found in the results of the tree datasets, the results presented in the following are obtained as mean values of those of the three cases.

Table 1 shows the positioning error results (root mean square error (RMSE)) obtained by using the device with uncalibrated sensors. More specifically, Table 1 compares the results obtained with the proposed revised particle filter presented in Section 3 with those obtained with the particle filter of Subsection 2.1. The particle filter of Subsection 2.1 has been run with $n = 100$ (in (a) and (d)), $n = 200$ (in (b) and (e)) and $n = 500$ (in (c) and (f)). Furthermore, two cases for the step lengths used in the particle filter have been considered: the length estimated by the sensor measurements (in (a), (b) and (c)), and the measured length corrected by the k factor (in (d), (e) and (f)). The results shown in Table 1 are the performance improvement (in percent) obtained by using the particle filter presented in Section 3 (with $n = 100$) with respect to that of Subsection 2.1. Since the results obtained by means of the particle filter are non-deterministic (*i.e.*, they depend on the specific realization of the random noise used to run the filter), the reported results have been obtained by averaging the results of 100 independent Monte Carlo simulations.

The standard deviations σ_s and σ_α have been set to proper values accordingly to our simulations: $\sigma_s = 0.15$ m and $\sigma_\alpha = \pi/6$.

Then, Table 2 shows the results obtained on the same track considered in Table 1, but using a (roughly) calibrated magnetic sensor. Calibration of the magnetic sensor has been performed as described in Section 4 by using 300 magnetic sensor sample measurements, $\{y_{m,t}\}_{t=1,...,300}$. The parameter values considered in Table 2 are the same considered in Table 1. Similarly to Table 1, the reported results have been obtained by averaging the results of 100 independent Monte Carlo simulations.

Table 1. Performance improvement (on the RMS positioning error) by using the revised particle filter with the uncalibrated magnetic sensor.

(a) $n = 100$	(b) $n = 200$	(c) $n = 500$	(d) $n = 100, ks_t$	(e) $n = 200, ks_t$	(f) $n = 500, ks_t$
51.6%	55.2%	55.2%	42.7%	40.0%	55.3%

Table 2. Performance improvement (on the RMS positioning error) by using the revised particle filter with the calibrated magnetic sensor.

(a) $n = 100$	(b) $n = 200$	(c) $n = 500$	(d) $n = 100, ks_t$	(e) $n = 200, ks_t$	(f) $n = 500, ks_t$
65.9%	62.1%	58.5%	45.5%	35.9%	33.6%

Figure 3 shows the magnetic sensor measurements and the corresponding fitting ellipse obtained with the calibration procedure described in Section 4. Calibration results are shown varying the number of sensor measurements used in the calibration process (sensor measurements are shown as cyan x-marks in Figure 3): the results obtained with 150 (top-left, solid green line), 200 (top-right, dashed blue line) and 300 (bottom-left, magenta dashed–dotted line) measurements are compared in the bottom-right sub-figure with those obtained with 1000 measurements (black stars) and with the fitting ellipse obtained with a ground truth calibration (red solid bold line). The fitting ellipse in the ground truth calibration corresponds to that of the (u_s, v_s) plane, which, during the ground truth calibration process, was coincident with the horizontal plane.

Some considerations are now in order. First, notice that the results of Table 2 have been obtained by using a roughly calibrated sensor, i.e., calibrated by using 300 sensor measurements. As shown in Figure 3 (bottom-right), 300 sensor measurements do not allow one to obtain a perfect calibration. However, since, in practice, the assumption of maintaining the (u_s, v_s) plane of a handhold device in an horizontal position can be only approximately satisfied, even with more measurements, the calibration results will typically differ from the ground truth ones: indeed, as shown in the bottom-right sub-figure of Figure 3, the fitting ellipse obtained by means of 300 and 1000 measurements are quite similar, but they both differ from the ground truth fitting ellipse (red solid bold line).

Nevertheless, actually, the use of a roughly calibrated magnetic sensor can allow one to improve the performance of the tracking system: Figure 4 shows an example of unconstrained trajectories (i.e., obtained by setting $\alpha_{i,t} = \alpha_t$, $s_{i,t} = s_t$ and without imposing the building physical constraints), where it is possible to compare the results obtained by using the uncalibrated magnetic sensor (blue x-marks), the magnetic sensor calibrated with 300 measurements (red x-marks, left sub-figure) and the magnetic sensor with the ground truth calibration (red x-marks, right sub-figure). As shown in the figure, the trajectory obtained with the magnetic sensor calibrated with 300 measurements is quite similar to the one obtained with the ground truth calibrated sensor. Since the trajectories shown in Figure 4 have been obtained by setting $\alpha_{i,t} = \alpha_t$, $s_{i,t} = s_t$, then they can be considered as (first) approximations of the average trajectories resulting from the particle filter (without imposing the building physical constraints) in the corresponding considered conditions: in this sense, the results shown in Figure 4 can be considered as representative of the typical outcomes. This is confirmed by the results of our

Monte Carlo simulations (Tables 1 and 2), where the use of the proposed online calibration allows one to reduce by 30%, approximately, the estimation error of both the proposed algorithm and the original particle filter.

Figure 3. Calibration of the magnetic sensor. Number of measurements used in the calibration: (**a**) 150, (**b**) 200, (**c**) 300. (**d**) Comparison of calibration results obtained with 150 (solid green line), 200 (dashed blue line), 300 (magenta dashed-dotted line), 1000 (black stars) measurements, with ground truth calibration (red solid bold line).

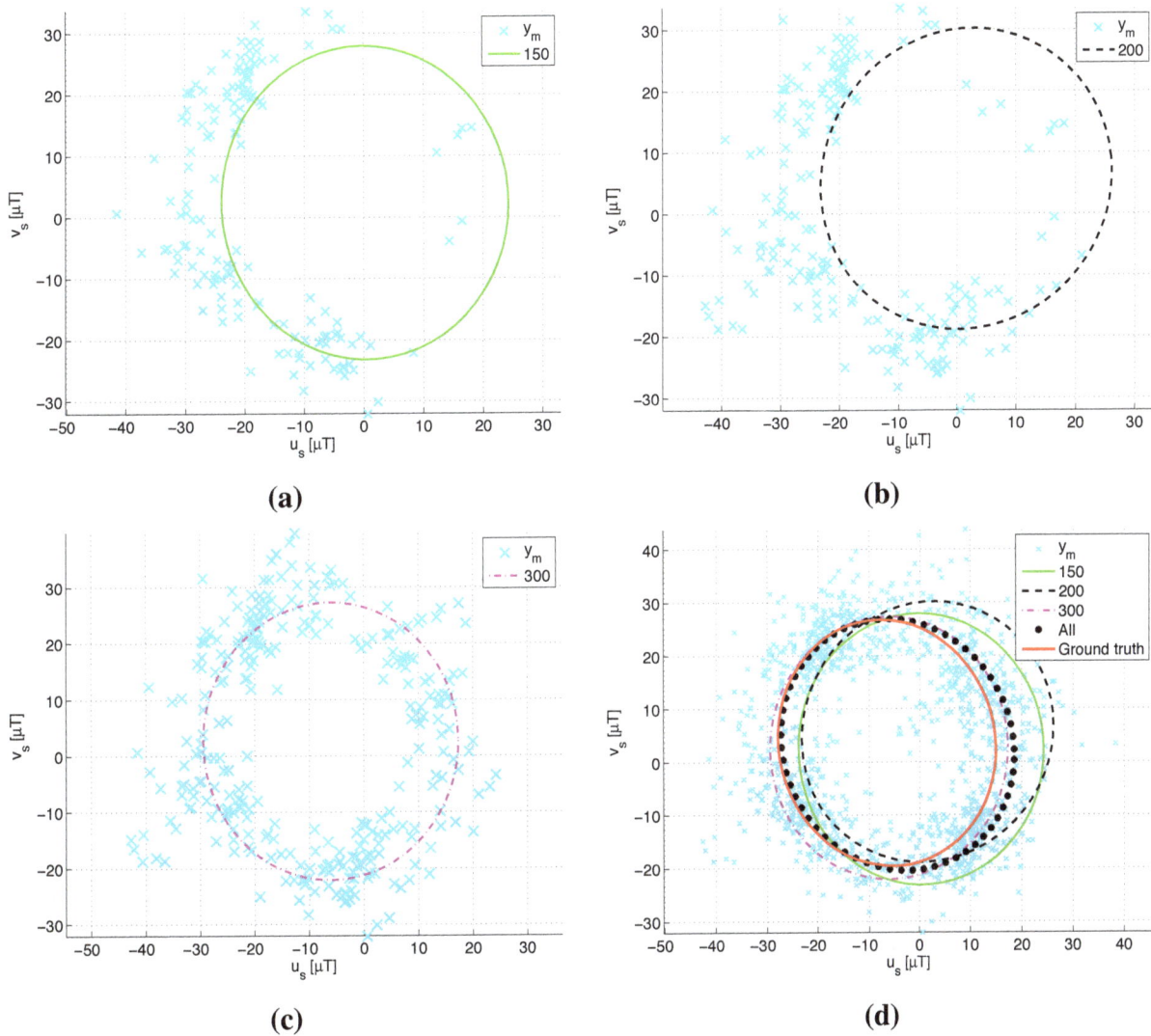

(a)

(b)

(c)

(d)

The results of Tables 1 and 2 show that the use of step lengths corrected by the factor k allows one to significantly reduce the tracking error of the particle filter. Furthermore, in all of the considered conditions, the revised particle filter proposed in Section 3 outperforms the other considered filtering approaches: despite being more clear in the uncalibrated case, its convenience is significant in the calibrated case, as well. This is probably due to the correction of local changes of the magnetic field (which yield local heading direction biases) that can be obtained by means of the proposed algorithm.

As expected, in the calibrated case, the improvement achieved by using the proposed algorithm decreases when n becomes larger; however, it is still significant even when the number of particles considered in the original particle filter (Subsection 2.1) becomes much larger than that of the revised particle filter of Section 3. In the uncalibrated case, since measurement errors are correlated, the original particle filter (Subsection 2.1) does not significantly improve increasing the number of particles from $n = 100$ to $n = 500$.

Figure 4. Comparison of unconstrained trajectories: trajectory obtained using uncalibrated (blue x-marks) and calibrated magnetic sensor (red x-marks). Different calibration models are considered: computed by using 300 magnetic sensor measurements (**left**), and ground truth sensor model (**right**).

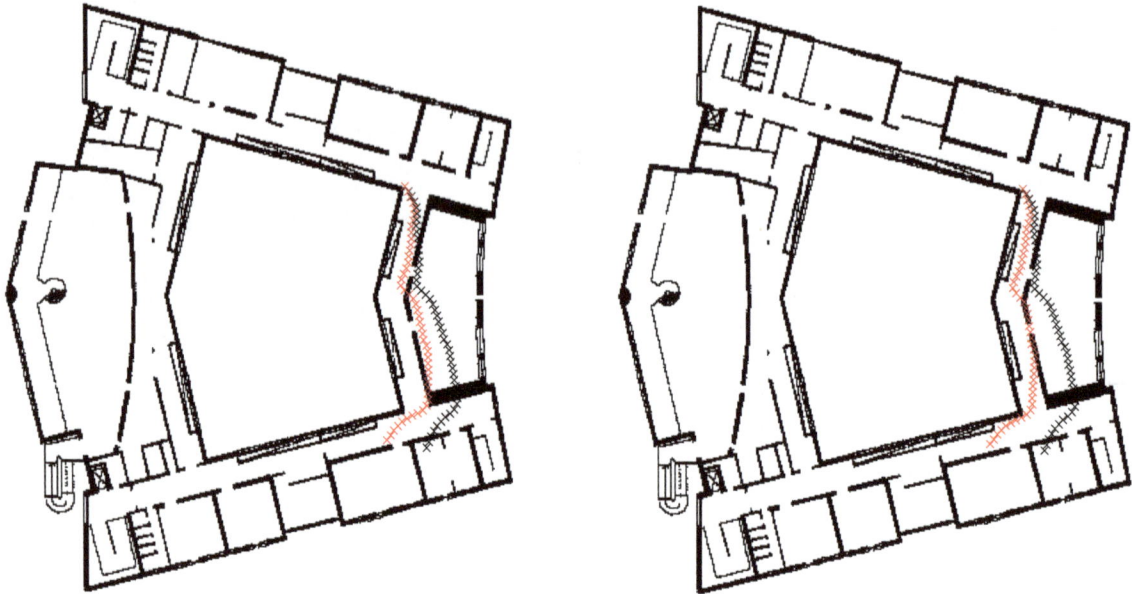

After the online calibration, the proposed algorithm achieves an average RMSE of (approximately) 3.5 m in the presented simulations by using $n = 100$ particles. Such performance is expected to improve by increasing the number of particles. However, considering that a larger number of n would significantly increase the computational load of the algorithm, which, according to our simulations, is approximately 10%–20% larger than in the particle filter of Subsection 2.1 with the same number of particles, $n = 100$ (however, notice that the computational load changes at each run of the algorithm, e.g., depending on the number of violated constraints).

According to the results obtained in the considered simulations, the particle filter approach proposed in Section 3 allows one to significantly improve the positioning results obtained with the filter of Subsection 2.1 in the case of uncalibrated embedded sensors and mobile device carried by the user's hand (notice that in [17,34], the use of the particle filter described in Subsection 2.1 was proposed for a different system, where sensors where integrated in the user's shoe). Such performance improvement has been obtained with a relatively small increase of the computational burden of the filter. Better results can be obtained by considering a larger number of particles n; however, issues with the real-time implementation of the system will arise for $n = \bar{n}$, with $\bar{n}$ sufficiently large. The actual value of $\bar{n}$ depends on the computational capabilities of the considered device ($n \approx 100$ particles can be used for real-time applications in standard mobile phones).

Notice also that despite the proposed approach allowing the use of a mobile device carried by the user's hand, the obtained results might be improved by imposing more restrictions on the system (e.g., fixing the device to the user body, for instance to the shoe, as in [17]), or by integrating other sensors (e.g., the integration of camera measurements has been considered in [14]). Furthermore, the considered approach allows a partial, but not complete, freedom of movement to the user's hand.

Motivated by the above considerations, our future works will aim to improve the proposed method, while preserving the use of a relatively small number of particles:

- The presented approach considers two different working conditions (*i.e.*, before and after the magnetic sensor calibration results are available). However, even if the magnetometer calibration results are not yet available, the data collected can be used to improve the algorithm presented in Section 3, e.g., in Section 3, the guessed bias angles $\hat{\alpha}_{b,t}$ and $\hat{\alpha}_{b,t'}$ are chosen independently; instead, it is clear that they are (partially) related accordingly to the sensor model: the selection of the bias angle $\hat{\alpha}_{b,t'}$ should be done accordingly to the data already collected.
- An extension of the proposed approach should be considered in order to let the heading direction freely vary with respect to the local coordinate system.

Finally, the presented navigation algorithm will be integrated in a mobile mapping system in order to provide accurate 3D photogrammetric reconstructions of the environment ([40–45]) based on the use of the camera embedded in the mobile device.

6. Conclusions

In this paper, a method for smartphone-based navigation has been proposed. The considered method is based on a particle filtering tracking approach and on the fusion of information from different uncalibrated sensors.

The proposed approach has been tested, with encouraging indoor navigation results, inside a building of the university: the considered algorithm outperforms a previously proposed approach, both in uncalibrated and in (partially) calibrated working conditions.

Some future work is planned to allow further improvement of the navigation ability of the proposed strategy and reducing the assumptions on the working conditions.

Author Contributions

All authors contributed equally in this work.

Conflicts of Interest

The authors declare no conflict of interest.

References

1. Bahl, P.; Padmanabhan, V. RADAR: An in-building RF-based user location and tracking system. In Proceedings of the INFOCOM 2000 Nineteenth Annual Joint Conference of the IEEE Computer and Communications Societies, Tel Aviv, Israel, 26–30 March 2000; Volume 2, pp. 775–784.

2. Casari, P.; Castellani, A.; Cenedese, A.; Lora, C.; Rossi, M.; Schenato, L.; Zorzi, M. The "Wireless Sensor Networks for City-Wide Ambient Intelligence (WISE-WAI)" Project. *Sensors* **2009**, *9*, 4056–4082.

3. Cenedese, A.; Ortolan, G.; Bertinato, M. Low-density wireless sensor networks for localization and tracking in critical environments. *Veh. Technol. IEEE Trans.* **2010**, *59*, 2951–2962.

4. Chang, N.; Rashidzadeh, R.; Ahmadi, M. Robust indoor positioning using differential Wi-Fi access points. *IEEE Trans. Consum. Electron.* **2010**, *56*, 1860–1867.

5. Chiou, Y.; Wang, C.; Yeh, S. An adaptive location estimator using tracking algorithms for indoor WLANs. *Wirel. Netw.* **2010**, *16*, 1987–2012.

6. Mazuelas, S.; Bahillo, A.; Lorenzo, R.; Fernandez, P.; Lago, F.; Garcia, E.; Blas, J.; Abril, E. Robust Indoor Positioning Provided by Real-Time. RSSI Values in Unmodified WLAN Networks. *IEEE J. Sel. Top. Signal Process.* **2009**, *3*, 821–831.

7. Youssef, M.; Agrawala, A. The horus WLAN location determination system. In Proceedings of the 3rd International Conference on Mobile Systems, Applications, and Services, Seattle, WA, USA, 6–8 June 2005; pp. 205–218.

8. Foxlin, E. Pedestrian tracking with shoe-mounted inertial sensors. *IEEE Comput. Graph. Appl.* **2005**, *25*, 38–46.

9. Ruiz, A.; Granja, F.; Prieto Honorato, J.; Rosas, J. Accurate Pedestrian Indoor Navigation by Tightly Coupling Foot-Mounted IMU and RFID Measurements. *IEEE Trans. Instrum. Meas.* **2012**, *61*, 178–189.

10. Barbarella, M.; Gandolfi, S.; Meffe, A.; Burchi, A. A Test Field for Mobile Mapping System: Design, Set Up and First Test Results. In Proceeding of the 7th International Symposium on Mobile Mapping Technology, Cracow, Poland, 13–16 June 2011.

11. Barbarella, M.; Gandolfi, S.; Meffe, A.; Burchi, A. Improvement of an MMS trajectory, in presence of GPS outage, using virtual positions. In Proceedings of the ION GNSS 24th International Technical Meeting of the Satellite Division, Portland, OR, USA, 20–23 September 2011; pp. 1012–1018.

12. Piras, M.; Marucco, G.; Charqane, K. Statistical analysis of different low cost GPS receivers for indoor and outdoor positioning. In Proceedings of the IEEE Position Location and Navigation Symposium (PLANS), Indian Wells, CA, USA, 4–6 May 2010; pp. 838–849.

13. Piras, M.; Cina, A. Indoor positioning using low cost GPS receivers: Tests and statistical analyses. In Proceedings of the 2010 International Conference on Indoor Positioning and Indoor Navigation (IPIN 2010), Zurich, Switzerland, 15–17 September 2010.

14. Saeedi, S.; Moussa, A.; El-Sheimy, N. Context-Aware Personal Navigation Using Embedded Sensor Fusion in Smartphones. *Sensors* **2014**, *14*, 5742–5767.

15. El-Sheimy, N.; Kai-wei, C.; Noureldin, A. The Utilization of Artificial Neural Networks for Multisensor System Integration in Navigation and Positioning Instruments. *IEEE Trans. Instrum. Meas.* **2006**, *55*, 1606–1615.

16. Lukianto, C.; Sternberg, H. Stepping–Smartphone-based portable pedestrian indoor navigation. *Arch. Photogramm. Cartogr. Remote Sens.* **2011**, *22*, 311–323.

17. Widyawan; Pirkl, G.; Munaretto, D.; Fischer, C.; An, C.; Lukowicz, P.; Klepal, M.; Timm-Giel, A.; Widmer, J.; Pesch, D.; *et al.* Virtual lifeline: Multimodal sensor data fusion for robust navigation in unknown environments. *Pervasive Mob. Comput.* **2012**, *8*, 388–401.

18. Masiero, A.; Guarnieri, A.; Vettore, A.; Pirotti, F. An indoor navigation approach for low-cost devices. In Proceedings of the 4th International Conference on Indoor Positioning and Indoor Navigation (IPIN 2013), Montbéliard, France, 28–31 October 2013.

19. Gordon, N.; Salmond, D.; Smith, A. Novel approach to nonlinear/non-Gaussian Bayesian state estimation. *IEE Proc. F Radar Signal Process.* **1993**, *140*, 107–113.

20. Doucet, A.; Johansen, A. *A Tutorial on Particle Filtering and Smoothing: Fifteen Years Later*; Technical Report; Department of Statistics, University of British Columbia: Vancouver/Kelowna, BC, Canada, 2008.

21. Doucet, A.; de Freitas, N.; Gordon, N. *Sequential Monte Carlo Methods in Practice*; Springer-Verlag: New York, NY, USA, 2001.

22. Zhang, Q.; Campillo, F.; Cerou, F.; Legland, F. Nonlinear system fault detection and isolation based on bootstrap particle filters. In Proceedings of the 44th IEEE Conference on Decision and Control, 2005 and 2005 European Control Conference (CDC-ECC '05) , Plaza de España Seville, Spain, 12–15 December 2005; pp. 3821–3826.

23. Dellaert, F.; Fox, D.; Burgard, W.; Thrun, S. Monte Carlo localization for mobile robots. In Proceedings of the 1999 IEEE International Conference on Robotics and Automation, Detroit, MI, USA, 10–15 May 1999; Volume 2, pp. 1322–1328.

24. Andrieu, C.; Davy, M.; Doucet, A. Efficient particle filtering for jump Markov systems. Application to time-varying autoregressions. *IEEE Trans. Signal Process.* **2003**, *51*, 1762–1770.

25. Guarnieri, A.; Pirotti, F.; Vettore, A. Low-cost mems sensors and vision system for motion and position estimation of a scooter. *Sensors* **2013**, *13*, 1510–1522.

26. Hol, J. Sensor Fusion and Calibration of Inertial Sensors, Vision, Ultra-Wideband and GPS. Ph.D. Thesis, The Institute of Technology, Linköping University, Linköping, Sweden, 2011.

27. Dorveaux, E.; Vissiére, D.; Martin, A.; Petit, N. Iterative calibration method for inertial and magnetic sensors. In Proceedings of the 48th IEEE Conference on Decision and Control, Shanghai, China, 15–18 December 2009; pp. 8296–8303.

28. Bonnet, S.; Bassompierre, C.; Godin, C.; Lesecq, S.; Barraud, A. Calibration methods for inertial and magnetic sensors. *Sens. Actuators A Phys.* **2009**, *156*, 302–311.

29. Schiavone, G.; Desmulliez, M.; Walton, A. Integrated Magnetic MEMS Relays: Status of the Technology. *Micromachines* **2014**, *5*, 622–653.

30. Kuo, J.T.W.; Yu, L.; Meng, E. Micromachined Thermal Flow Sensors—A Review. *Micromachines* **2012**, *3*, 550–573.

31. Hautefeuille, M.; O'Flynn, B.; Peters, F.H.; O'Mahony, C. Development of a Microelectromechanical System (MEMS)-Based Multisensor Platform for Environmental Monitoring. *Micromachines* **2011**, *2*, 410–430.

32. Au, A.K.; Lai, H.; Utela, B.R.; Folch, A. Microvalves and Micropumps for BioMEMS. *Micromachines* **2011**, *2*, 179–220.

33. Jahn, J.; Batzer, U.; Seitz, J.; Patino-Studencka, L.; Gutiérrez Boronat, J. Comparison and evaluation of acceleration based step length estimators for handheld devices. In Proceedings of the 2010 International Conference on Indoor Positioning and Indoor Navigation (IPIN), Zurich, Switzerland, 15–17 September 2010; pp. 1–6.

34. Widyawan; Klepal, M.; Beauregard, S. A Backtracking Particle Filter for Fusing Building Plans with PDR Displacement Estimates. In Proceedings of the 5th Workshop on Positioning, Navigation and Communication (WPNC 2008), Hannover, Germany, 27 March 2008; pp. 207–212.

35. Douc, R.; Cappe, O. Comparison of resampling schemes for particle filtering. In Proceedings of the 4th International Symposium on Image and Signal Processing and Analysis (ISPA 2005), Southhampton, UK, 15–17 September 2005; pp. 64–69.

36. Vasconcelos, J.; Elkaim, G.; Silvestre, C.; Oliveira, P.; Cardeira, B. Geometric Approach to Strapdown Magnetometer Calibration in Sensor Frame. *IEEE Trans. Aerosp. Electron. Syst.* **2011**, *47*, 1293–1306.

37. Liu, Y.; Li, X.; Zhang, X.; Feng, Y. Novel Calibration Algorithm for a Three-Axis Strapdown Magnetometer. *Sensors* **2014**, *14*, 8485–8504.

38. Golub, G.; Loan, C.V. *Matrix Computations*; Johns Hopkins University Press: Baltimore, MD, USA, 1989.

39. Gower, J.; Dijksterhuis, G. *Procrustes Problems*; Oxford University Press: New York, NY, USA, 2004.

40. Hartley, R.; Zisserman, A. *Multiple View Geometry in Computer Vision*; Cambridge University Press: Cambridge, UK, 2003.

41. Ma, Y.; Soatto, S.; Kosecka, J.; Sastry, S. *An Invitation to 3-D Vision: From Images to Geometric Models*; Interdisciplinary Applied Mathematics: Imaging, Vision, and Graphics Volume 26; Springer: New York, NY, USA, 2004.

42. Agarwal, S.; Snavely, N.; Seitz, S.; Szeliski, R. Bundle Adjustment in the Large. In Proceedings of the European Conference on Computer Vision, Lecture Notes in Computer Science, Crete, Greece, 5–11 September 2010; Volume 6312, pp. 29–42.

43. Masiero, A.; Guarnieri, A.; Vettore, A.; Pirotti, F. An ISVD-based Euclidian structure from motion for smartphones. *ISPRS—Int. Arch. Photogramm. Remote Sens. Spat. Inf. Sci.* **2014**, *XL-5*, 401–406.

44. Forlani, G.; Giussani, A.; Scaioni, M.; Vassena, G. Target detection and epipolar geometry for image orientation in close-range photogrammetry. *ISPRS—Int. Arch. Photogramm. Remote Sens.* **2012**, *XXXI-5*, 518–523.

45. Feng, T.; Liu, X.; Scaioni, M.; Lin, X.; Li, R. Real-time landslide monitoring using close-range stereo image sequences analysis. In Proceedings of the 2012 International Conference on Systems and Informatics (ICSAI), Yantai, China, 19–20 May 2012; pp. 249–253.

Appendix

If the (initial) heading direction with respect to the local coordinate system is not conventionally fixed *a priori*, it can be estimated from the first steps of the user, as described in the following. First, the system is assumed to be switched on while the device is stationary and oriented according to its ideal orientation of use with respect to the (initial) heading direction: the value of the gravitational acceleration g is measured during this phase. The first K user's steps ($K = 2$ in the simulations of Section 5) are assumed to be taken along a straight line, and the device is kept in the ideal orientation of use: then, ideally the accelerations detected by the sensor (in an inertial system) should lie on a plane, that has a simple basis formed by g and the unit vector corresponding to the heading direction. In practical operating conditions, in order to reduce the effect of measurement noise (and other non-ideal conditions), such a plane is estimated as that generated by the first two principal directions of a principal component analysis of the measured accelerations. Then, from the above considerations, the heading direction can be easily computed from the values of g and of the basis of the plane. Finally, the sign of the heading direction (*i.e.*, if the user is going forward or backward along that direction) can be easily determined by the sign of the accelerations projected along that direction.

3

Insulin Micropump with Embedded Pressure Sensors for Failure Detection and Delivery of Accurate Monitoring

Dimitry Dumont-Fillon, Hassen Tahriou, Christophe Conan and Eric Chappel *

Debiotech SA, 28 avenue de Sévelin, 1004 Lausanne, Switzerland;
E-Mails: d.dumont-fillon@debiotech.com (D.D.-F.); h.tahriou@debiotech.com (H.T.);
c.conan@debiotech.com (C.C.)

* Author to whom correspondence should be addressed; E-Mail: e.chappel@debiotech.com

External Editor: Joost Lötters

Abstract: Improved glycemic control with insulin pump therapy in patients with type 1 diabetes mellitus has shown gradual reductions in nephropathy and retinopathy. More recently, the emerging concept of the artificial pancreas, comprising an insulin pump coupled to a continuous glucose meter and a control algorithm, would become the next major breakthrough in diabetes care. The patient safety and the efficiency of the therapy are directly derived from the delivery accuracy of rapid-acting insulin. For this purpose, a specific precision-oriented design of micropump has been built. The device, made of a stack of three silicon wafers, comprises two check valves and a pumping membrane that is actuated against stop limiters by a piezo actuator. Two membranes comprising piezoresistive strain gauges have been implemented to measure the pressure in the pumping chamber and at the outlet of the pump. Their high sensitivity makes possible the monitoring of the pumping accuracy with a tolerance of ±5% for each individual stroke of 200 nL. The capability of these sensors to monitor priming, reservoir overpressure, reservoir emptying, outlet occlusion and valve leakage has also been studied.

Keywords: MEMS; micropump; piezoresistive pressure sensor; failure detection; insulin delivery

1. Introduction

Traditional syringe pumps deliver insulin through tubing. These devices are reported to be safe and reliable, despite several drawbacks related to tubing kinking and to the so-called stick-slip effect of the plunger, an inherent limitation affecting short-term accuracy. Alternative systems for pumping small fluid volumes have been produced over the last three decades [1,2]. Among these various kinds of devices, the reciprocating displacement micropump based on MEMS technology and having two check valves together with a fixed stroke volume has shown the capability to combine both low cost and good performance [3,4].

An improved MEMS-based micropump comprising two integrated pressure sensors has been engineered in order to accurately deliver insulin to patients diagnosed with type 1 and 2 diabetes mellitus. The micropump delivers short-acting insulin 24 h a day through a cannula placed under the skin. Switching from multiple daily infusion (MDI) to continuous subcutaneous insulin infusion (CSII) induces well-known advantages, including the reduction of swings in blood glucose levels and the limitation of severe hypoglycemia [5–7]. In addition, this MEMS-based micropump provides outstanding delivery accuracy and failure detection capabilities, as reported in recent *in vitro* and *in vivo* studies comparing the pumping capability of several pump systems [8]. In this paper, the emphasis is placed on the description of the integrated pressure sensors and their typical experimental responses in various conditions. The implementation of a pressure sensor in the fluidic line is a well-known method to detect the occlusion of syringe pumps. Geipel *et al.* have reported a microfluidic system, including a commercial pressure sensor placed downstream of the pump to monitor the pulsed infused flow [9]. The two main challenges of these dynamic pressure measurements in a catheter are the control of the dead volume and the elasticity of the fluidic line. The use of pressure profiles to detect catheter occlusions and disconnects of a drug delivery system comprising notably a pressurized reservoir and microvalves with embedded pressure sensors has been also reported by Li *et al.* [10]. We show here that the integration of a pressure sensor within the pumping chamber is a powerful tool to understand the fluid dynamics inside the device. Thanks to its high compression ratio and small elasticity, this micropump is able to detect any malfunction (bubble, leakage) that leads to an accuracy loss larger than 5%. In contrast to syringe pumps, which control the residual volume of the insulin reservoir via the monitoring of the plunger position, microfluidic devices are usually not able to guarantee correct infusion, in particular in the case of failure. This integrated pressure sensor is therefore an essential element of the insulin micropump to prevent any severe hyperglycemic episodes.

The device shown in Figure 1 is a reciprocating displacement MEMS-based micropump having two check valves and a fixed stroke volume [11,12]. The pump chip is made of a stack of three plates: a top Si cap, an SOI (Silicon-On-Insulator) wafer and a bottom Si cap. The main elements of the micropump presented in Figure 1 are: the inlet valve, the pumping chamber, the pumping membrane having a mesa mechanically coupled to a piezo actuator (not represented here), the inner sensor, the outlet valve and the outer sensor.

Because the pumping membrane is located in-between the two mechanical stops when at rest, the micropump is called a "push-pull" device. Due to its high compression ratio, about 1.9:1 for a nominal stroke volume of 200 nL, the device exhibits self-priming capabilities. The piezo actuator is driven at ±200 V to push/pull the pumping membrane alternately against the two stop limiters. The actuation frequency in the bolus mode is fixed at 3.125 Hz. In the basal mode, the actuation profile is optimized to conciliate low power consumption and high delivery accuracy [12].

The filling of the pump is associated with a negative relative pressure in the pumping chamber, which opens the inlet valve and closes the outlet valve (the pull of the pumping membrane). Inversely, the infusion corresponds to positive relative pressure in the pumping chamber, which opens the outlet valve and closes the inlet valve (the push of the pumping membrane).

The silicon pump chip is attached to a ceramic substrate and linked to a piezo actuator, thus forming a Pumpcell. In order to complete the disposable unit, this element is set onto a fluidic block that comprises, in particular, the insulin reservoir, a hydrophilic particle filter, a battery and an outlet needle that is inserted into the cannula port during the placement of the pump onto the patch. A reusable electronics (pump controller) is finally assembled by the patient onto this disposable unit before the filling. Pictures of the device are provided in Figure 2.

Figure 1. Schematic cross-section of the micropump (not to scale). The arrow indicates the flow direction. The small rectangles are schematic representations of the anti-bonding structures. The green layer is the buried oxide.

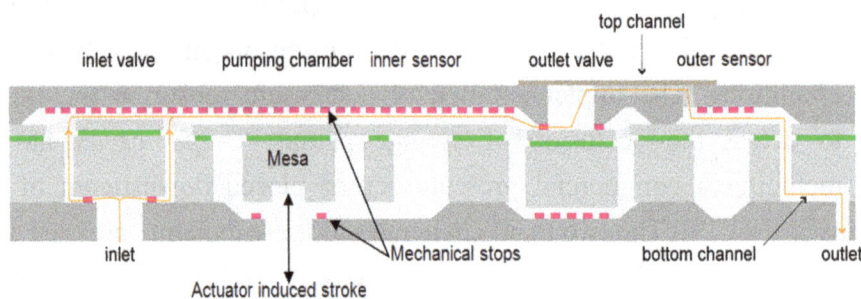

Figure 2. Photos of the device (courtesy of Debiotech SA, Switzerland). (**a**) Pumpcell comprising the micropump mounted onto a ceramic substrate having interconnection pads. (**b**) Disposable unit with the Pumpcell and the battery (top) mounted onto the insulin reservoir (bottom). (**c**) Pumping unit ready to be attached onto the patch. The pump controller (top) is assembled with the disposable unit (bottom). (**d**) Pumping unit and its remote controller.

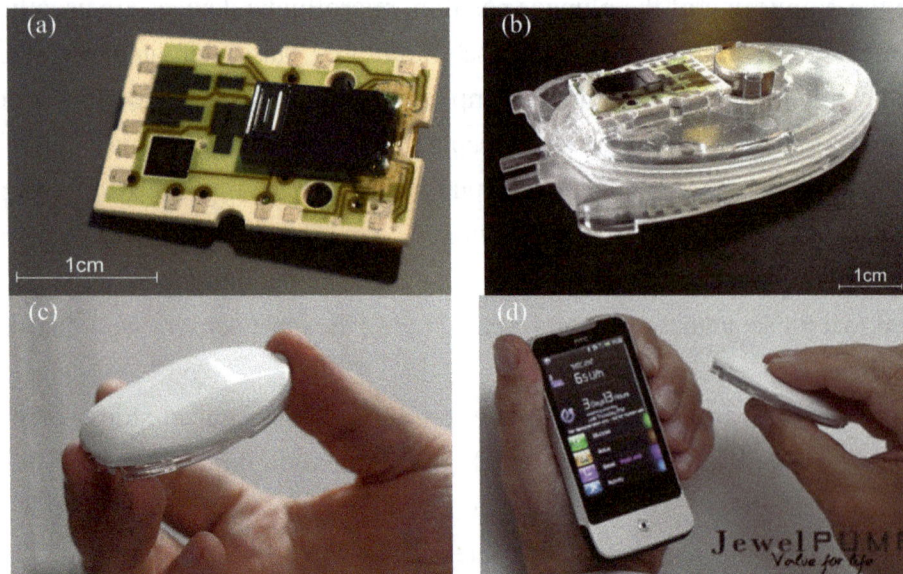

MEMS technology is particularly adapted to the integration of piezoresistive pressure sensors in a silicon chip. The first membrane having strain gages laid out in a Wheatstone bridge configuration is placed in the pumping chamber to monitor the pump dynamics, while the second one is placed downstream of the outlet valve for improved occlusion detection.

Resistors aligned along [110] crystallographic directions are obtained by boron implantation of an n-doped Si membrane built from an SOI wafer (100). The sheet resistance of this p-doped layer is about 1500 ohms per square. Each resistor is made of four squares. The sensor membrane is a square of about 1 mm^2, with a thickness of 15 μm. The physical characteristics of these gauge pressure sensors have been optimized to get a maximum offset of ± 2 mV·V^{-1}·bar^{-1} and a sensitivity of 18 mV·V^{-1}·bar^{-1} in the range -1 to $+1.5$ bar. The minimum resolution is 1 mbar.

2. Experimental Section

The micropumps are characterized in a clean room using the precision scale Sartorius MC5 (microgram resolution). A glass capillary glued to the pump cannula is immersed into a glass beaker prefilled with insulin Novorapid (from Novo Nordisk) and placed onto the scale. A thin layer of paraffin oil is used to prevent evaporation during the test. The temperature is regulated at 20 ± 1 °C for scale stabilization purposes. The pressure sensor signals were recorded with data acquisition cards NI 6143 (National Instruments, Austin, TX, USA). A pressure controller, Druck DPI520 (GE Sensing, Billerica, MA, USA), is used to simulate occlusion at the outlet, reservoir overpressure or an empty reservoir. The micropump is placed in an oven for tests at 38 °C in order to simulate the highest in-use temperature conditions for a patch pump. The relative position between the micropump and the meniscus of the glass beaker is adjusted before each experiment. The stabilization of the scale is the key parameter to obtain reproducible results. The scale is placed on an anti-vibration table. Specific care has been taken during the handling of the beaker, the introduction of the capillary through the paraffin oil layer and the removal of air bubbles inside the beaker, as well as the fluidic line. If not specified, the reservoir is filled with fast acting insulin Novorapid.

3. Results and Discussion

3.1. Priming

The priming frequency is fixed at 3.125 Hz. Typical sensor signals are provided in Figure 3. Before the first water introduction, the pressure in the pumping chamber is driven by the valve pretensions, while the pressure peaks at the pump outlet remain very small. Both signals could be used to monitor the pump priming, since the change of the signal shape after the introduction of liquid in the pumping chamber is significant. The height increase of the pressure peak in the pumping chamber during priming is mainly related to a change of compressibility, while the height of the pressure peak at the outer sensor location is more related to the viscosity change (between gas and liquid). Numerical models of the fluid dynamics in the pump have been built to evaluate the best tradeoff between pumping efficiency and failure detectability.

Figure 3. Inner and outer sensor signals during priming.

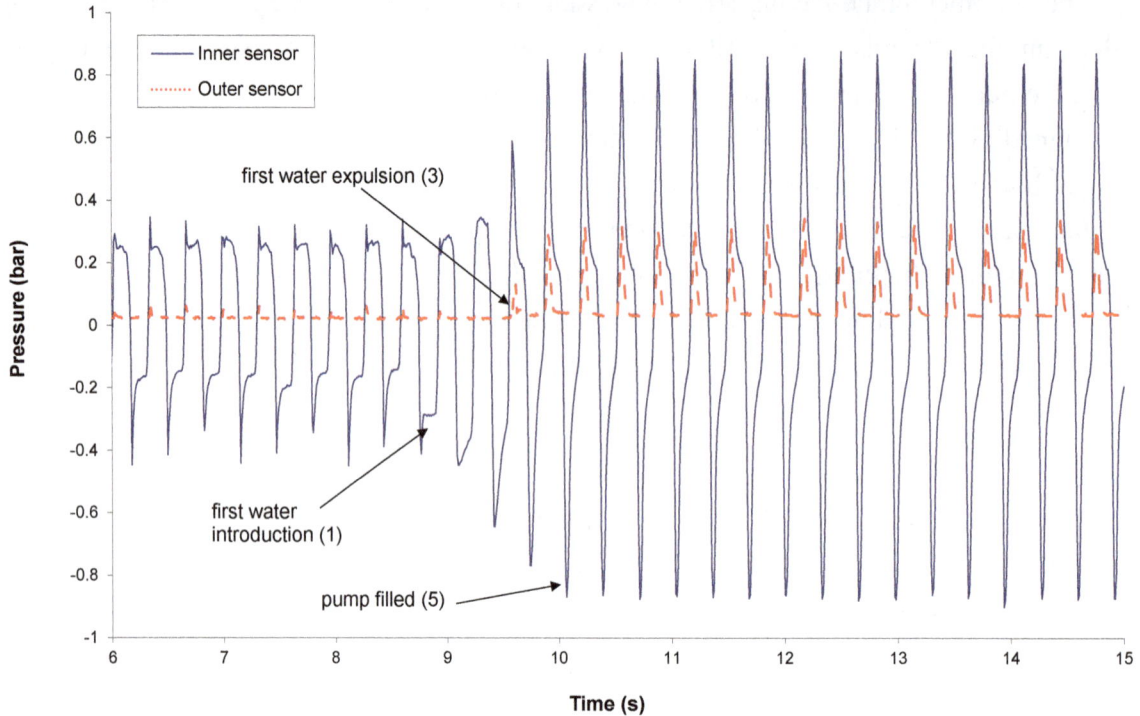

3.2. Occlusion

Occlusion is a common failure of insulin pumps. This is due to either a real obstruction caused by insulin fibrils at the outlet, a kinking of the cannula or a compression of the skin around the infusion site. An occlusion leads to two kinds of risk:

- Hyperglycemia: during the occlusion itself, since there is no infusion of insulin;
- Hypoglycemia: during the release of the occlusion (e.g., if the compression of the skin around the infusion site or the kinking of the cannula is suddenly released), there is a major risk of over-infusion, because the accumulated amount of insulin that has not been injected during the occlusion is injected very quickly.

The occlusion detection is therefore a key factor for patient safety.

Occlusion detection has been tested experimentally by monitoring the sensor signals before and after the occlusion of the pump cannula with clamp tweezers. The micropump is actuated in basal mode at 7.2 U/h (equivalent to 72 μL of insulin U 100 per hour or one stroke every 10 s). In Figures 4 and 5 are shown, respectively, the inner and outer sensor signals before the occlusion and during the five strokes following the occlusion. For the sake of clarity, the sensor signals are only shown during the actuation phase.

Because the fluidic pathway downstream the micropump exhibits a very low compressibility, any occlusion will induce a large increase of the outlet pressure, which could be detected by both sensors. The outer sensor, which is in direct communication with the infusion site (*i.e.*, without any valve in-between), is preferably used to monitor occlusion. Total occlusion could be detected during the first actuation that follows the event, thus triggering an alarm and stopping the pump to prevent the generation of large pressure, which may damage the fluidic line.

Figure 4. Evolution of the inner sensor signal before and during the five strokes following an occlusion.

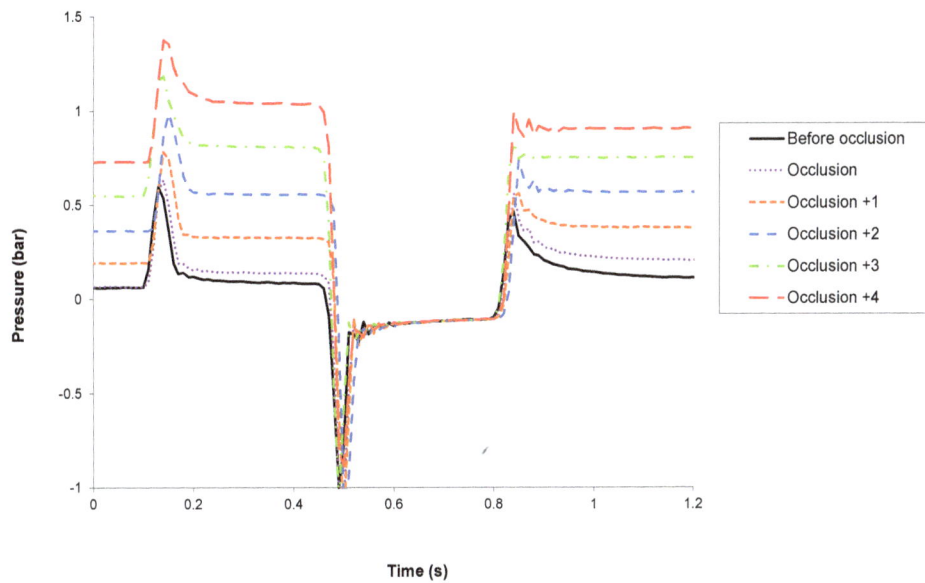

Figure 5. Evolution of the outer sensor signal before and during the five strokes following an occlusion.

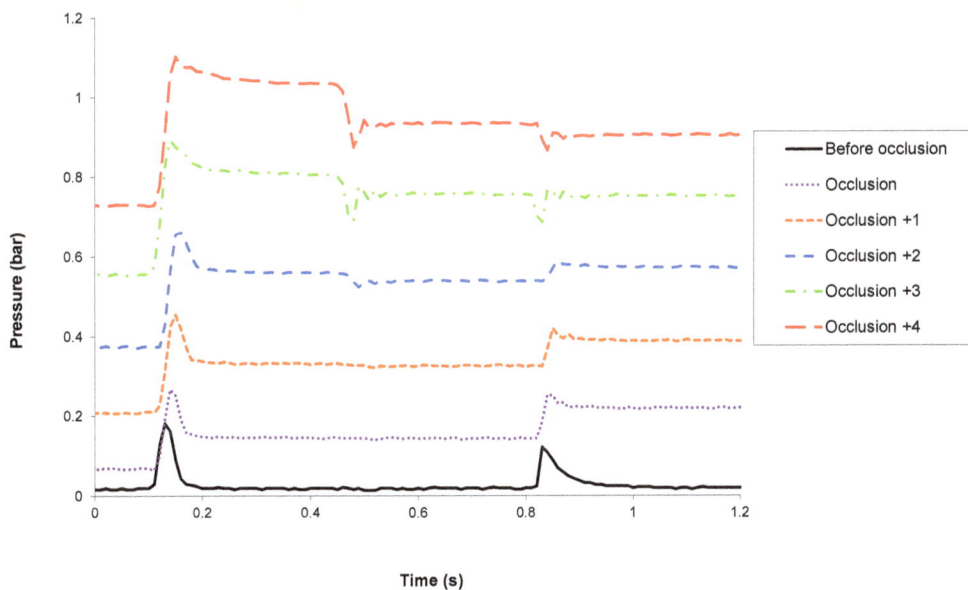

3.3. Reservoir Overpressure and Underpressure

The evolution of the stroke volume as a function of the inlet pressure has been studied from large negative pressure, corresponding to reservoir emptying, to large positive pressure. The latter condition can only occur in the case of reservoir overfilling, since the pump housing and reservoir structures prevent any transfer of pressure to the insulin of the reservoir.

Figure 6 shows that in both standard and reverse mode, the stroke volume remains constant in the range −0.3 bar to +0.2 bar. This is in good agreement with the check valve pretension values in the range of 0.1 bar. Standard (respectively reverse) mode refers to an actuation cycle that starts with a push (respectively pull).

Reservoir over- and under-pressures are monitored in the filling phase of the pumping cycle (*i.e.*, when the inlet valve opens during a pull), using the inner sensor. Typical pressure profiles in standard mode are provided in Figure 7, for inlet pressures varying from −0.3 to +0.3 bar.

Figure 6. Stroke volume as a function of the inlet pressure, in standard and reverse actuation modes.

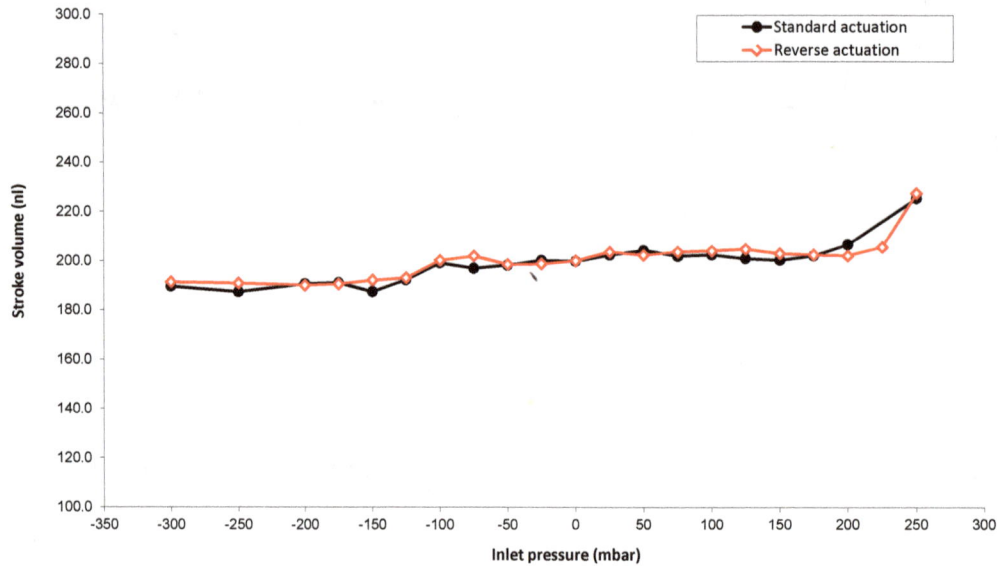

Figure 7. Inner sensor signal *versus* inlet pressure during an actuation cycle.

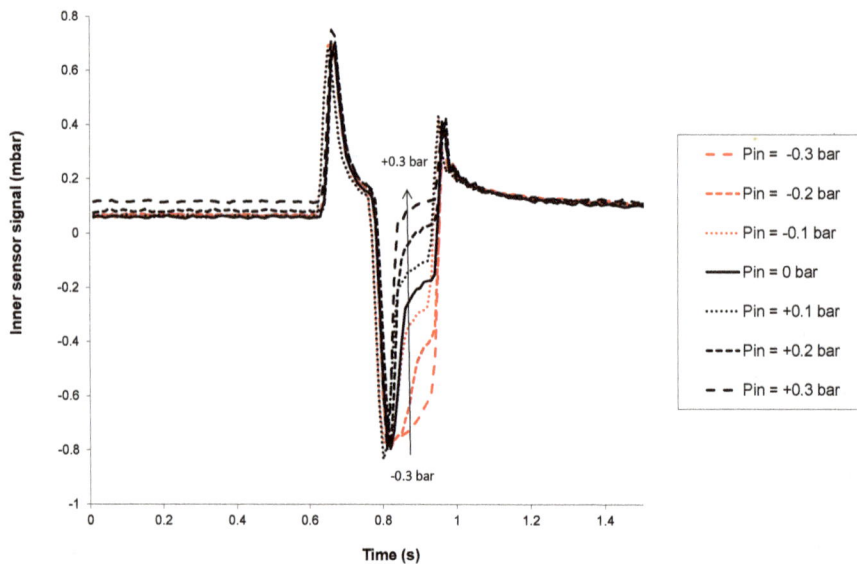

3.4. Stroke Volume and Actuation Cycle Characteristics

Figure 8 shows that the stroke volume does not vary with the insulin basal rate, from a slow basal rate of 0.1 U/h to a very high basal rate of 70 U/h.

The stroke volume as a function of the voltage driving the piezo actuator is provided in Figure 9. Under normal pressure conditions, the mechanical stops are reached at 100 V. The piezo is overdriven to 200 V in order to compensate for the assembly and machining tolerances, as well as different outer pressure conditions.

Figure 8. Stroke volume as a function of the flow rate in standard and reverse modes (for insulin U 100; 1 U/h is equivalent to 0.01 mL/h).

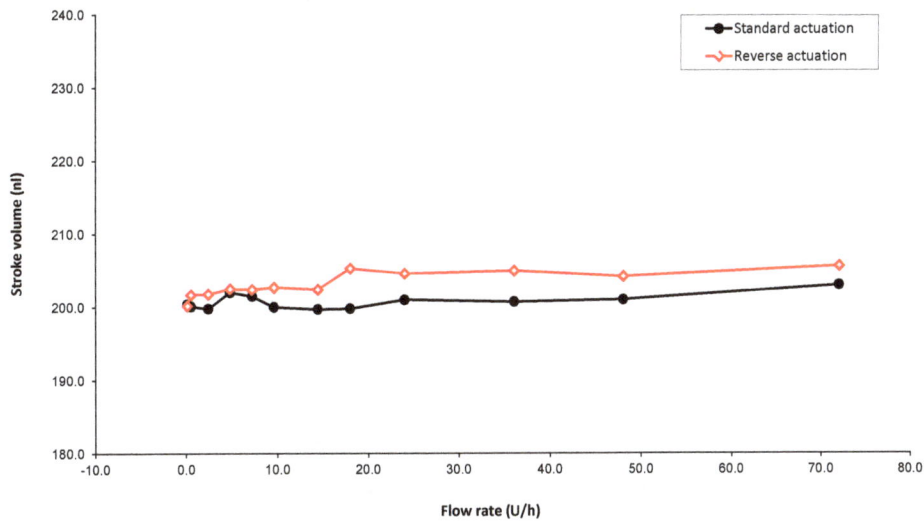

Figure 9. Stroke volume as a function of the actuation voltage in standard and reverse modes.

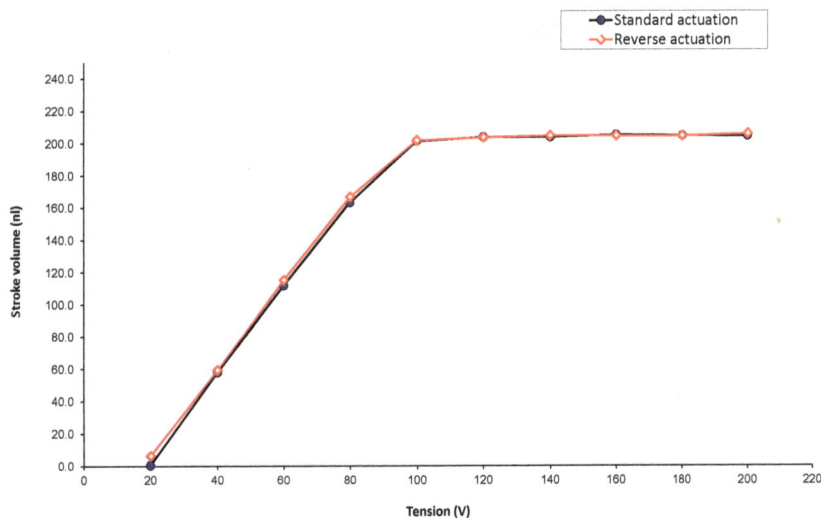

3.5. Stroke Volume and Viscosity

The stroke volume is independent of viscosity for incompressible media, since the device is a volumetric pump, but the maximum flow rate is reduced for viscous liquid.

The tests have been performed at 20 °C using water and a mix of water and glycerol. The actuation frequency has been increased up to the maximum flow rate. For water, the flow rate varies linearly with the actuation frequency up to 2.5 mL/h (actuation frequency of 3.5 Hz), and a maximum flow rate of 4.3 mL/h is measured at 18 Hz. The results are summarized in Table 1.

Table 1. Max flow rate as a function of viscosity.

Fraction of glycerol by mass	Viscosity (cP)	Max flow rate (mL/h)	Frequency (Hz)
0	1	4.3	18
0.59	10	0.85	6
0.71	25	0.46	6

3.6. Leakage Detection

The effect of a leakage of the pumping chamber, typically due to particles on the valve seat, is shown in Figure 10. The test has been performed by pumping a water solution containing 6×10^5 particles/mL (0.5 micron silica particles from Duke Scientific). The particle filter of the micropump has been removed for the test. Since the fluidic resistance of this filter is more than 100-times smaller than the fluidic resistance of the pump chip itself, this removal does not change the pump dynamics. The test has been carried out up to the measurement of a significant leakage. Infrared (IR) inspection has revealed an important amount of silica particles forming clusters on both valve seats.

It is worth noting that the pressure decay following the outlet valve closure, during the push phase, is significantly changed in the case of a back-flow equivalent to a decrease of 5% of the nominal stroke volume. The leakage detection algorithm is based on the analysis of the pressure decay just after the pressure peaks, which is compared to a threshold value being a fraction of the outlet valve pretension. Any accuracy loss larger than 5% would therefore be detected. In addition, the absence of positive pressure in the reservoir prevents any insulin flow in the case of valve leakage.

Figure 10. Inner sensor signal in reverse mode before and after a leakage, which reduces the effective stroke volume (SV) by about 5%.

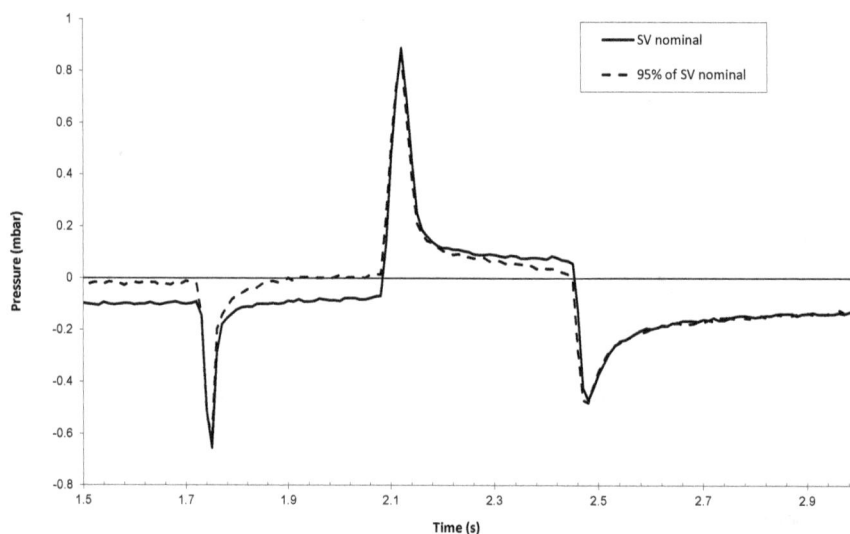

3.7. Air Bubble Detection

To investigate the effect of air bubbles on the inner sensor signal, the inlet port of the micropump has been connected to a reservoir with tubing. This tubing has been disconnected during one actuation cycle to introduce an air bubble of 200 nL in the fluidic line. The pump is then actuated in bolus mode (at a frequency of 3.125 Hz). Figure 11 shows typical pressure curves recorded by the inner sensor during the introduction of this bubble in the pumping chamber. The air is introduced during Stroke 1 in the pumping chamber, inducing an enlargement and a height reduction of the pressure peaks. The air bubble detection algorithm is therefore based on the analysis of the shape of the inner sensor pressure peaks. The air bubble in the pumping chamber is then expelled progressively during the following strokes. These results are only qualitative, because of the fractioning of the air bubble, making the estimation of the quantity of air in the pumping chamber during the test difficult.

Figure 11. Five consecutive pressure profiles of the inner sensor during the introduction of an air bubble of 200 nL in the pumping chamber. The air bubble is introduced during Stroke 1 and is progressively expelled during the three following strokes.

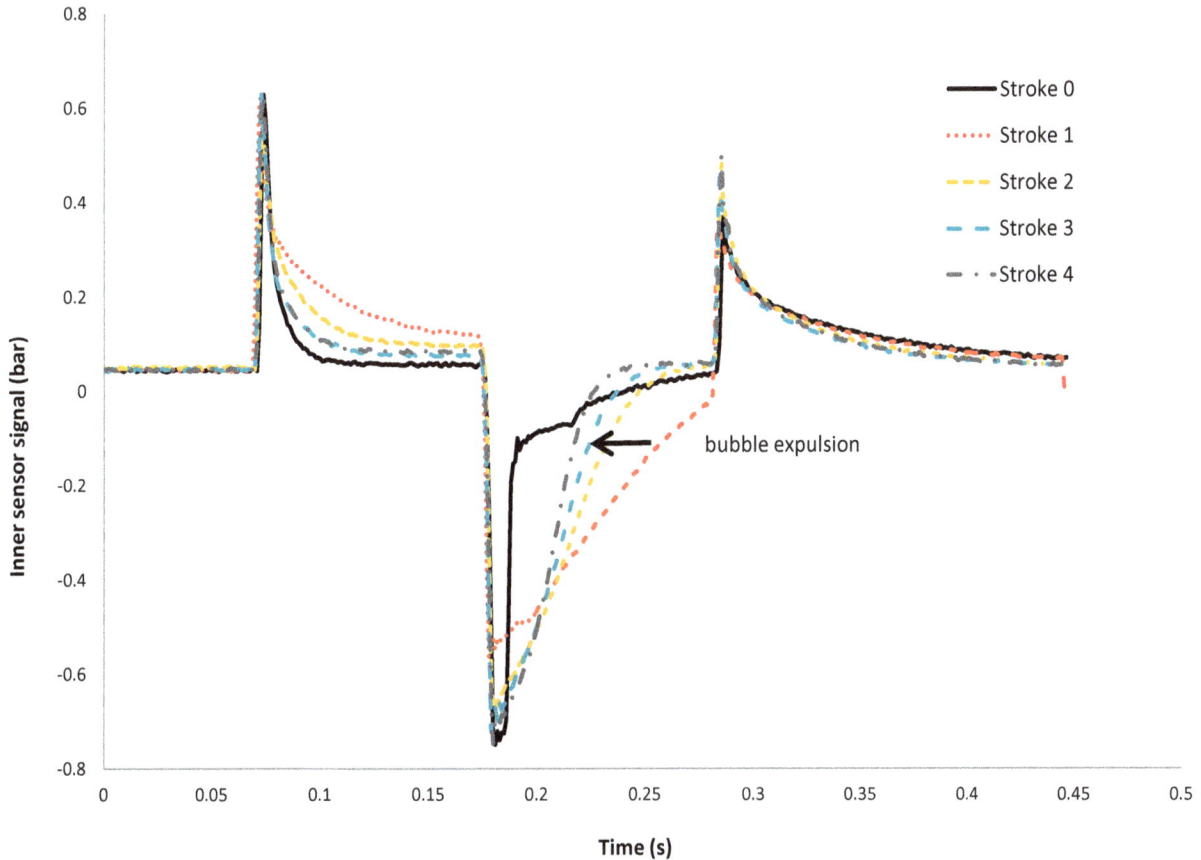

3.8. Delivery Accuracy

Pump accuracy is commonly illustrated by the so-called trumpet curves. The flow is measured by weighing the infused insulin in predefined observation windows. The pump is programmed to deliver insulin at a fixed rate. The flow data are processed to estimate the evolution of the maximum and the minimum percentages of deviation as a function of the observation windows defined according to the norm IEC60601-2-24 [13]. Data are recorded and processed with Infuscale 7.0 (Infuscale, Munich, Germany).

The document [12] shows trumpet curves obtained at room temperature, for a programmed flow rate of 2 U/h. In Figure 12 are provided trumpet curves obtained using a micropump programmed at 1 U/h, which corresponds to an actuation period of 72 s. Only the pumping unit has been placed into an oven at 38 °C. The curves have been extracted from data collected 1 h after the start of the infusion (the stabilization period fixed by the norm has been intentionally reduced from 24 h to 1 h). Despite the limitations induced by the test setup itself (in particular, the instabilities due to the temperature gradient in the flexible tube between the scale and the pumping unit), delivery accuracy better than ±5% is obtained for observation windows larger than four minutes. The mean flow error with respect to the set rate is equal to +0.28% here.

Figure 12. Maximum and minimum flow rate errors as a function of the observation window (*i.e.*, "trumpet curve" according to IEC60601-2-24). The dashed line is the mean % of error equal to +0.28%.

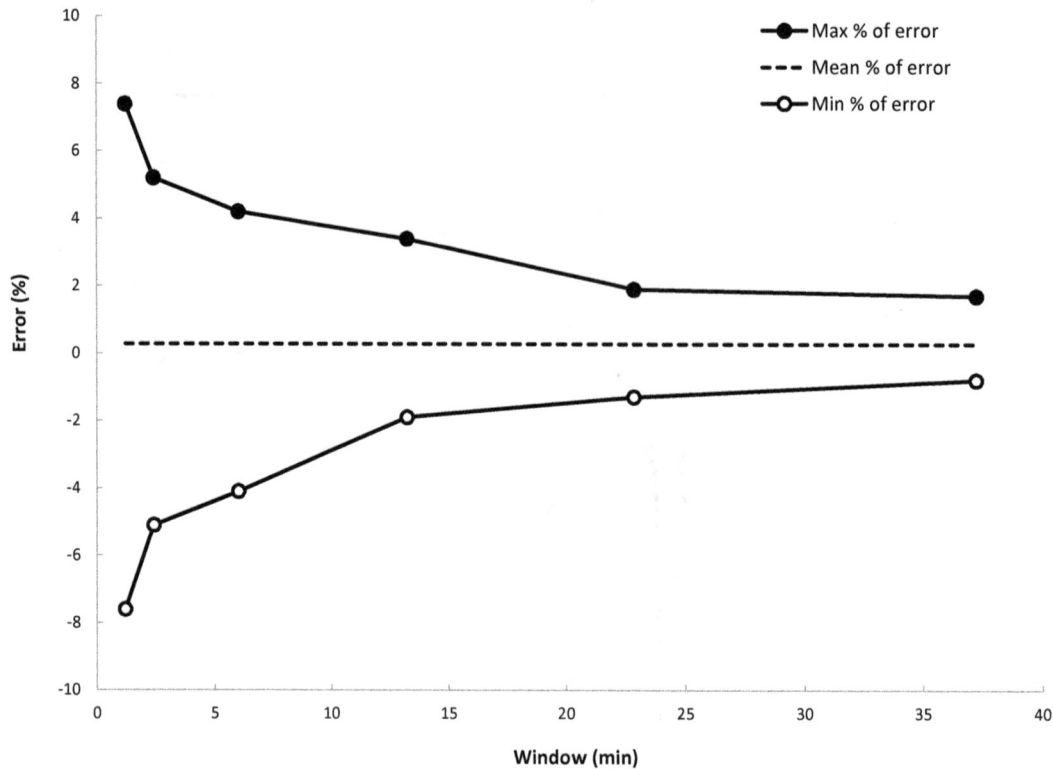

4. Conclusions

A MEMS-based micropump dedicated to insulin delivery has been characterized in various conditions of use. In addition to its capability to pump with very high accuracy up to a rate of 2.5 mL/h, the device comprises powerful integrated piezoresistive pressures sensors able to detect malfunctions, like occlusion, reservoir over/under pressure, air bubbles or leakage, which could lead to an accuracy loss larger than 5%. The presence of a sensor in the pumping chamber able to monitor delivery accuracy is a key element to prevent severe hypoglycemic episodes and to improve the safety of future closed-loop systems.

Acknowledgments

The authors would like to thank the whole team involved in the Insulin Micropump Project at Debiotech and ST Microelectronics.

Author Contributions

Eric Chappel wrote the paper and led the design activities with Dimitry Dumont-Fillon. Hassen Tahriou and Christophe Conan performed the characterization tests.

Conflicts of Interest

The authors declare no conflict of interest.

References

1. Woias, P. Micropumps-past, progress and future prospects. *Sens. Actuators B* **2005**, *105*, 28–38.
2. Laser, D.J.; Santiago, J.G. A review of micropumps. *J. Micromech. Microeng.* **2004**, *14*, 35–64.
3. Van Lintel, H.T.G.; van de Pol, F.C.M.; Bouwstra, S. A piezoelectric micropump based on micromachining of silicon. *Sens. Actuators* **1988**, *15*, 153–167.
4. Maillefer, D.; van Lintel, H.; Rey-Mermet, G.; Hirschi, R. A high-performance silicon micropump for an implantable drug delivery system. In Proceedings of the 12th IEEE International Conference on Micro Electro Mechanical Systems, Orlando, FL, USA, 17–21 January 1999.
5. Doyle, E.A.; Weinzimer, S.A.; Steffen, A.T.; Ahern, J.A.; Vincent, M.; Tamborlane, W.V. A randomized, prospective trial comparing the efficacy of continuous subcutaneous insulin infusion with multiple daily injections using insulin glargine. *Diabetes Care* **2004**, *27*, 1554–1558.
6. Hirsch, I.B.; Bode, B.W.; Garg, S.; Lane, W.S.; Sussman, A.; Hu, P.; Santiago, O.M.; Kolaczynski, J.W. Continuous subcutaneous insulin infusion (CSII) of insulin aspart *versus* multiple daily injection of insulin aspart/insulin glargine in type 1 diabetic patients previously treated with CSII. *Diabetes Care* **2005**, *28*, 533–538.
7. Jeitler, K.; Horvath, K.; Berghold, A.; Gratzer, T.W.; Neeser, K.; Pieber, T.R.; Siebenhofer, A. Continuous subcutaneous insulin infusion *versus* multiple daily insulin injections in patients with diabetes mellitus: Systematic review and meta-analysis. *Diabetologia* **2008**, *51*, 941–951.
8. Borot, S.; Franc, S.; Cristante, J.; Penfornis, A.; Benhamou, P.Y.; Guerci, B.; Hanaire, H.; Renard, E.; Reznik, Y.; Simon, C.; Charpentier, G. Accuracy of a new patch pump based on a microelectromechanical system (MEMS) compared to other commercially available insulin pumps: Results of the first *in vitro* and *in vivo* studies. *J. Diabetes Sci. Technol.* **2014**, *8*, 1133–1141.
9. Geipel, A.; Goldschmidtboeing, F.; Jantscheff, P.; Esser, N.; Massing, U.; Woias, P. Design of an implantable active microport system for patient specific drug release. *J. Biomed. Microdevices.* **2008**, *10*, 469–478.
10. Li, T.; Evans, A.T.; Chiravuri, S.; Gianchandani, Y.B. Compact, power-efficient architectures using microvalves and microsensors, for intrathecal, insulin, and other drug delivery systems. *Adv. Drug Deliv. Rev.* **2012**, *64*, 1639–1649.
11. Chappel, E.; Allendes, R.; Bianchi, F.; Calcaterra, G.; Cannehan, F.; Conan, C.; Lefrique, J.; Lettieri, G.L.; Mefti, S.; Noth, A.; Proennecke, S.; Neftel, F. Industrialized functional test for insulin micropumps. In Proceedings of the 25th Eurosensors Conference, Athens, Greece, 4–7 September 2011.
12. Chappel, E.; Mefti, S.; Lettieri, G.L.; Proennecke, S.; Conan, C. High precision innovative micropump for artificial pancreas. In Proceedings of the 12th Microfluidic, BioMEMS and Medical Microsystems Conference, San Francisco, CA, USA, 1–6 February 2014.
13. International Electrotechnical Commission. IEC60601-2-24, 1st ed.; IEC: Geneva, Switzerland, 1998.

4

Generation of Nanoliter Droplets on Demand at Hundred-Hz Frequencies

Slawomir Jakiela, Pawel R. Debski, Bogdan Dabrowski and Piotr Garstecki *

Institute of Physical Chemistry, Polish Academy of Sciences, Kasprzaka 44/52, 01-224 Warsaw, Poland;
E-Mails: sjakiela@ichf.edu.pl (S.J.); pdebski@ichf.edu.pl (P.R.D.); b.dabrowski86@gmail.com (B.D.)

* Author to whom correspondence should be addressed; E-Mail: garst@ichf.edu.pl

External Editor: Joost Lötters

Abstract: We describe a precision micropump for generation of precisely metered micro-aliquots of liquid at high rates. The use of custom designed piezoelectric valves positioned externally to the microfluidic chip allows for on-demand formation of micro-droplets with online control of their individual volumes from nLs to μLs at frequencies up to 400 Hz. The system offers precision of administering volumes of 1% and of time of emission of <0.5 ms. The use of a piezoelectric actuator provides two distinct vistas for controlling the volume of the droplets—either by digital control of the "open" interval or by analogue tuning of the lumen of the valve. Fast and precise generation of droplets make this system a perfect constituent module for microfluidic high-speed combinatorial screening schemes.

Keywords: microfluidics; droplets on demand; nanoliter droplets; piezoelectric valve

1. Introduction

Droplet microfluidic techniques support almost all kinds of reactions in chemistry, biochemistry, and microbiology with the most pronounced advantages brought in the screening of large random libraries [1–3]. The most important and outstanding challenges in construction of multiphase microfluidic systems include automation of the operations on individually addressable microdroplets. Preferably, such individual control over the timing of flow, volumes and content of drops should be exerted with high precision and in a possibly inexpensive format, preferably, with disposable chips that minimize the cost of operation and the risks of cross-contamination between experiments. Lab-on-Chip systems that allow for fast and, in particular, precise generation of droplets of predefined volume and composition may find a wide variety of applications in biochemistry, especially in high-throughput screening. For example, they can be used for high-throughput screening and bioassays. Also, they will make it simple to run digital tests, *i.e.*, digital polymerase chain reaction (PCR) [4–6] and digital enzyme-linked immunosorbent assay (ELISA) [6,7], in Lab-on-Chip systems, as they allow to generate large numbers of droplets of precisely defined volume and composition.

Churski *et al.* [8] proposed the use of *external* electromagnetic valves for the technical simplicity of the approach and for its compatibility with single-layer chips fabricated in virtually any material. The disadvantage of the electromagnetic valves used in the demonstration of the system [8] and in an integrated system for studying epistatic interactions between antibiotics [9] is the limited rate of operation of the coil actuators. One straightforward way to alleviate this problem is to use the much faster piezoelectric actuators as they offer high speed and precise reproducibility of motion. Piezoelectric actuators *integrated* with microfluidic chips have been used for ejection [10], sorting [11] and generation [12–15] of droplets.

There are a number of reports on generation of droplets on demand in microfluidic systems. The approaches include both integrated actuators [12–14] and modular designs [8,16] that lower the complexity of the microfluidic chips and allow for their easy exchange. Piezoelectric actuators have been shown to generate droplets with high precision (coefficient of variation (CV) ~0.3% [12]). The range of volumes expressed as the ratio of the volume of largest droplet that can be generated to the smallest drop is yet typically small, less than 10 in the systems utilizing piezo-electric actuators [12–14]. Other schemes of actuation increase the dynamic range of volumes to above few tens [16,17] or even approximately one hundred [8]. Some of the solutions operate at high rates, for example the "pico-injectors" reported by Abate *et al.* [16] generate droplets at frequency of up to 10 kHz, yet the precision of these systems is small with coefficients of variation typically above 10% [16,17]. In summary, there is no demonstration to date of a system combining high precision of dosing, wide range of volumes and high rates of operation.

Here we report the construction and application of an external piezoelectric valve for the generation of micro-droplets on demand in a wide range of volumes with much higher accuracy than available in current solutions. We report on the speed and precision of the system and discuss the use of the valve in larger, more complex microfluidic systems. Our design, being external to the microfluidic chip, is not influenced by the choice of the chip material. Therefore it can be used not only for Polydimethylsiloxane (PDMS), but also for polycarbonate (PC), glass or silicon chips. Moreover, an external generator can be used multiple times in various systems, which lowers the cost of a single chemical reaction and the complexity of the experimental set-up. Furthermore, if there are two or more generators integrated with the chip, they can influence each other spoiling their performance. External generators are free from this problem. Such robustness of external generators, including the design presented here, is crucial for the reproducibility of biological and biochemical reactions that require very precise and accurate dosing of substrates.

We report a microfluidic system that can generate precisely metered droplets on demand at frequencies in excess of hundred Hz. The system uses custom piezo-electric valves positioned outside the microfluidic chip. The valves control the time of emission of each individual droplet with a precision of *ca.* 0.5 ms, and the volume of each droplet with a precision better than 10 pL within the whole 1 nL to 20 µL range (over 4 log) of volumes that can be administered into each drop. The system is compatible with a wide range of materials of which the chips are made and with various dimensions of the microchannels. We demonstrate both digital and analogue control of the volume of the droplets, and operation of the system paced at up to 400 Hz.

2. Experimental Section

2.1. Microfluidic Chip

In the tests of the range and precision of administering volumes of micro-droplets and of the rate of operation of the valve we used simple microfluidic chips comprising a T-junction [18,19]. We fabricated polycarbonate chips in 5-mm thick plates (Macroclear, Bayer, Leverkusen, Germany) using milling machine (MSG4025, Ergwind, Gdansk, Poland). The mean of absolute values of profile of milled surface of channel averaged over the profile equals 2.6 µm, and the maximum height of the profile was 4.8 µm (calculated with the use of Bruker CountourGT-K0 profilometer, Bruker, Billerica, MA, USA). The milled plates were thermally bonded to flat 2-mm plates by compressing them together (30 min, 130 °C). Two valves (Figure 1a) were supplied with liquids (one with hexadecane with 0.5% (*w/w*) Span 80, second with distilled water) from pressurized reservoirs. We connected the chip inlets to resistive steel capillaries (outer diameter (O.D.) 200 µm, inner diameter (I.D.) 100 µm, length 1–5 m, Mifam, Milanowek, Poland) extending from the valves using short segments of Tygon tubing (~2 cm, O.D. 2.01 mm, I.D. 0.19 mm, Ismatec, Glattbrugg, Switzerland) to connect the capillaries with the needles.

Figure 1. (**a**) Schematic diagram of the experimental system: each of the liquid supply lines comprises a pressurized reservoir, a valve, and a resistive capillary that interfaces the chip. (**b**) Scheme of the valve: A—the body of the valve, B—valve inlet, C—valve outlet (capillary tube), D—membrane, E—pushrod, F—piezoelectric actuator, G—brass-steel housing.

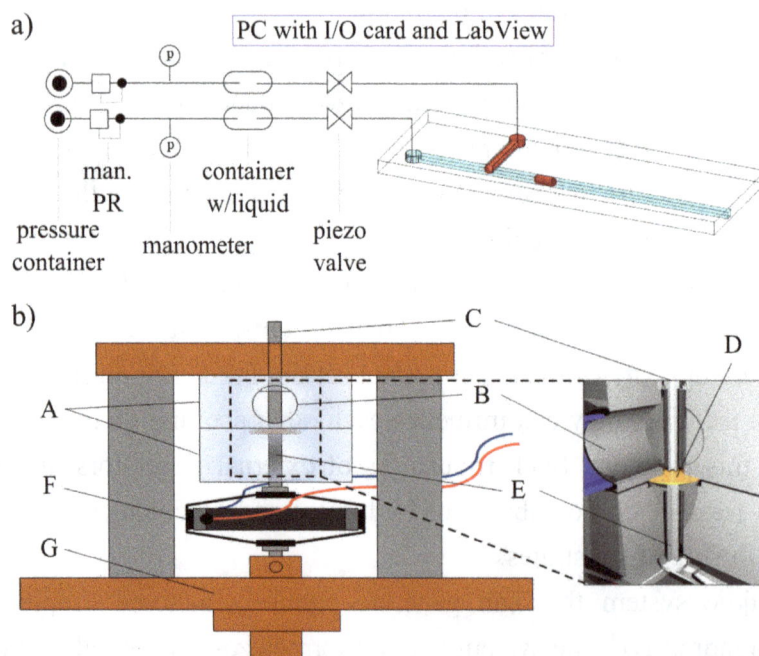

2.2. The Valve

We used APA40SM (Cedrat, Meylan, France, capacity of 1.8 µF) piezoelectric actuators, nominally providing up to 52 µm stroke, up to 4 kHz frequency (resonance frequency is 4.1 kHz without load), and up to 194 N of force. We milled the body of the valve (marked A in Figure 1, see also Figure S1 in supplementary) in a block of polycarbonate (PC). In the outlet of the body of the valve we placed a stainless steel capillary (C) of O.D. 1.3 mm and I.D. 1.1 mm and 2.5 cm length. The liquid was supplied to the valve (B) from a pressurized reservoir via a wide tubing (O.D. 6 mm, I.D. 4 mm). We used a polyester-laminated aluminium foil membrane (D) (150–110 µm thick aluminium foil and 20 µm thick polyester layer on both sides, Heinz Herenz Gmbh, Hamburg, Germany) to close the outlet of the valve. The membrane was pressed against the inlet into the steel capillary with a steel pushrod (E in Figure 1) of a diameter of 2 mm and a length of 15 mm, connected directly to the piezoelectric actuator (F in Figure 1). Both the valve and the actuator were mounted in a brass-steel housing (G) that allowed for fine tuning of the distance between the two parts with the use of a detachable micrometric screw (Thorlabs, Newton, NJ, USA). The actuator in the normal (extended) state was pressing the membrane down against the steel capillary, while applying electric potential withdrew the steel rod and allowed for a finite lumen between the edge of the steel capillary and the membrane. The valve provided for a displacement of up to 50 µm at frequencies of up to 400 Hz. Above 400 Hz, the inertia of the steel pushrod decreased the amplitude of motion of the actuator (Figure S2 in supplementary).

2.3. Amplifier

To control the piezoelectric actuator, we built a custom proportional voltage amplifier characterized by a 15× gain of the potential for 0–10 V input at frequencies from 0 to 4 kHz [20]. The amplifier allows us to control the piezoelectric actuator digitally (on/off) and with the pulse width modulation (PWM) with the duty cycle d (percent of time that the potential is on) ranging from 10% to 90% (Figure S3 in supplementary). One of the most important parameters of the voltage amplifier is the maximum current that it can supply to the actuator. Higher current allows us to control an actuator with larger capacitance for the same working frequency. Our amplifier supports the 3µF actuator at its full range of frequencies, $i.e.$, up to 4 kHz. In our experiments we used the National Instrument (Austin, TX, USA) cards (PCIE-6321) that generated synchronized PWM signals for two amplifiers to ensure the proper control of the valves. Custom written LabView script enabled the change of duty cycle for each step. Our solution can be multiplied up to 24 independent amplifiers connected to valves.

2.4. Protocol of Generation of Droplets on Demand

A microfluidic droplet on demand (DOD) system should allow us to produce droplets of an arbitrary ($i.e.$, within a possibly wide range) volume and at an arbitrary time of emission. In order to achieve a wide dynamic range of accessible volumes we actively control both immiscible phases [8]. In order to generate a droplet of required volume, the LabView protocol: (i) closes the flow of the continuous (oily) phase, (ii) opens the flow of the droplet (aqueous) phase and keeps it open for an interval t_{open}, (iii) closes the flow of the droplet phase, and (iv) opens the flow of the continuous liquid to break-off the droplet and push it downstream of the junction. A well-performing DOD system should show an ideally linear relation between the interval t_{open} and the volume V of the droplet.

Our system enables to connect a number of different sources of the flow with usually different chemical factors. The only condition for proper operation of a complex system is that each source of flow has a higher hydrodynamic resistance in the steel capillary than the hydrodynamic resistance of the microchannels on the chip and simultaneously has the comparable resistance with other sources. This is achieved with the use of long thin capillaries. We have described the rules for building these systems in Churski *et al.* [21], and shown examples of complex systems operated within this technology [22].

2.5. Measurement of the Volume of Droplets

We imaged the droplets with a fast camera (Photron PCI1024, Photron, Tokyo, Japan) through a microscope (Nikon Eclipse E200, Nikon, Tokyo, Japan). We measured the lengths of the droplets in a tube (I.D. 0.2 mm, polytetrafluoroethylene (PTFE), Bola, Grünsfeld, Germany) connected to a square cross-section microchannel and modeled their volume as a cylinder caped with two hemispheres. This model shape of the droplet may introduce small systematic error to the estimate of the volume, yet it does not compromise the measurement of the precision and reproducibility of the volumes of droplets generated on demand in our system. The applied technique allows us to measure the volume with the precision of 0.1 pL, which was limited by our ability to optically resolve the interfaces and the deformation of ends of droplets.

3. Results and Discussion

3.1. Tuning the Volume of Droplets via Timing the Valves

We first tested the ability of our system to control the volume of the droplets formed by changing the value of t_{open} while keeping the constant electric potential (150 V) applied to the actuator. We recorded the volumes of droplets generated by our system at five different values of the pressure applied to the reservoir of water p_{water}. In all these experiments the pressure p_{oil} was constant (1500 mbar), resulting in the instantaneous rate of flow of oil being always 25 mL/h. We observed that (i) the volume of droplets increases linearly with increasing t_{open} (Figure 2) and (ii) the rate of this increase is proportional to p_{water}. These results show that the rate of flow of the droplet phase can be well described with the Hagen-Poiseuille law ($Q \sim \frac{\Delta p}{R}$, where $Q = \frac{dV}{dt}$ is volumetric flow rate, Δp is the pressure drop and R is the length of capillary) for a viscous flow in a capillary (inset of Figure 2 shows relationship: $\frac{\Delta V}{\Delta p/R} \sim \Delta t$).

3.2. Tuning the Volume of Droplets Generated at a Constant Frequency

The system can be used to tune the volume of monodisperse droplets created at a constant frequency. If we set the duration of a single tact of the system as $\tau = 1/f$, the interval t_{open} when the valve for the droplet phase is open can be expressed as $t_{open} = d/f$. We call $d = t_{open}/\tau$ the duty cycle in analogy to the pulse width modulation technique (Figure S3). With the use of our setup we were able to tune d in the range of 0.1 to 0.9. This limits the dynamic range Ω of volumes to $\Omega = \frac{V_{d=0.9}}{V_{d=0.1}} \sim 9$. This is well reflected in our experimental results (Figure 3a). As we increase the frequency (decrease τ), the range of droplets that can be generated shifts to smaller volumes ($V \sim t_{open} = d/f$), yet the ratio of the largest and smallest achievable droplets stays approximately constant ($\frac{V_{d=0.9}}{V_{d=0.1}} \sim \frac{\frac{0.9}{f}}{\frac{0.1}{f}} = 9$ inset of Figure 3a).

Figure 2. The graph shows the volumes of droplets as a function of the interval t_{open} and of the pressures p_{water} applied to the container with water. The inset shows almost perfect (99.6% level of compliance) agreement of the measured trends with the Hagen-Poiseuille law (see inset). The droplets were formed in 400×400 μm^2 square cross-section channel.

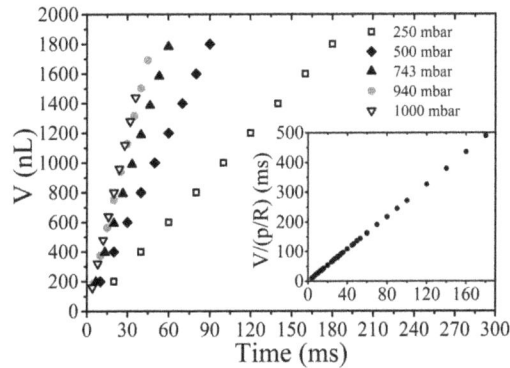

Figure 3. Formation of drops of different volumes in a single cycle obtained in (**a**) digital control and (**b**) analogue control. Insets show the ratio of maximum to minimum volume of drops as a function of the frequency of valve operation. The droplets were formed in 100×100 μm^2 square cross-section channel. During the experiments the values of pressure p_{oil} (1500 mbar) and p_{water} (2000 mbar) were constant.

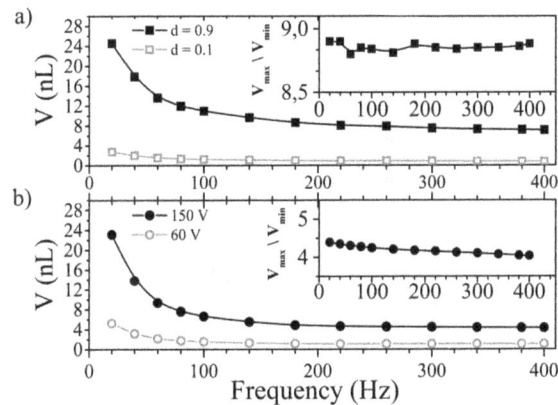

3.3. Analogue Control

Similar function of tuning the volume of monodisperse drops produced at a fixed frequency can be achieved via the *analogue* control of the potential applied to the piezoelectric actuator. We fix the duty cycle d to 0.5 and change the voltage applied to the valve between 60 and 150 V. This corresponds to a range of 20 to 50 μm micrometers of deflection of the actuator (Figure S4 in supplementary). Figure 3b shows the dependence of the maximum and minimum volume that could be generated for a given frequency via this method. As one can see, the dynamic range of achievable volumes slightly decreases with increasing frequency (inset of Figure 3b).

3.4. Maximum Frequency of Operation

We quantified the range of frequencies at which droplets can be formed for a fixed set of parameters of the system (v = 200 mm/s—average velocity of the oil in the microchannel, d = 0.5, U = 150 V). The system operated correctly at frequencies lower than 400 Hz. In this regime all droplets were monodisperse, the maximum relative error in administering the volumes, expressed as the standard deviation of the diameter normalized by the mean (for at least 20 consecutive drops), was less than 1% (Figure 4).

Figure 4. Graph shows the relative error of formed droplets as a function of their volumes. Each chart corresponds to appropriate size of square cross-section of channel. Droplets were formed at constant frequency 400 Hz and the pressure applied to the container with water (2500 mbar) was fixed in all experiments. In turn, pressure applied to the container with oil was 4000 mbar for 100×100 μm^2, 2400 mbar for 200×200 μm^2, and 2000 mbar for 400×400 μm^2, which resulted in constant volumetric flow of the oil in the microchannel in all instances. The smallest reproducibly (when relative error is lower than 1%) droplets are marked and they are comparable with the volume define by the cube of width of channel.

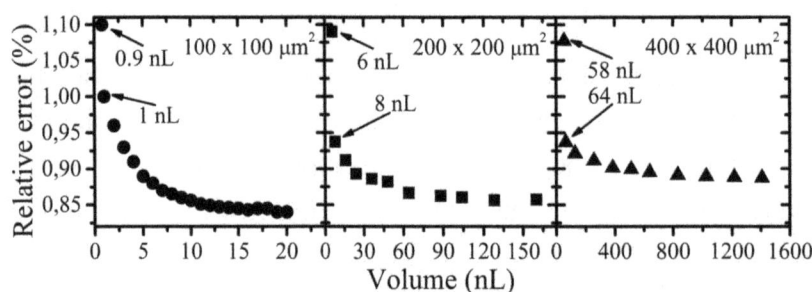

3.5. Generation of Arbitrary Sequences of Volumes of Droplets

We have also verified the ability of our system to generate sequences of droplets of varied, and arbitrarily predefined volume at a constant frequency (50 Hz—Figure 5, and 400 Hz—Figure S5 in supplementary) of operation. We first calibrated the relation between the volume V of the droplets and the duty cycle d:$V(d)$, as explained above. We then programmed a sequence of volumes V_i of droplets into the script that controlled the valves. The protocol listed different values of the duty cycle d_i in each (i-th) cycle of the operation of the system. We then run the script using LabVIEW (National Instruments, Austin, TX, USA) and recorded a video to analyze the actual volumes of droplets produced. We used a sequence that—when plotted for volume against index of the droplet in the sequence—shows the acronym of our institute ("ICHF" for "Instytut CHemii Fizycznej") (Figure 5). The apparent scatter ("bold font" of the letters) was intentionally programmed into the script. The measured volumes of droplets coincided within 99.8% with the programmed values. Figure S5 illustrates similar experiment in a smaller microchannel—i.e., 100×100 μm^2, in which droplets were formed with 400 Hz frequencies. Correlation coefficient is equal to 0.9991 in this instance.

Such sequence can be used in larger, integrated systems [9,22,23], to merge droplets of different chemical composition to form sequences of mixtures that systematically screen the composition space of a reaction.

Figure 5. (a) The graph shows the pre-programmed (dots) and measured (open circles) volumes of droplets produced at 50 Hz in our system. The droplets were formed in 400×400 μm^2 square cross-section channel. The scatter is intentional—o fill in the "body" of the letters coding for ICHF, the acronym of the Institute of Physical Chemistry (Instytut CHemii Fizycznej, Warszawa, Poland). (b) The graph illustrates the same sequence of droplets and shows the accuracy in forming of droplet on demand. Pearson correlation coefficient between measured and programmed volume of droplet is 0.998.

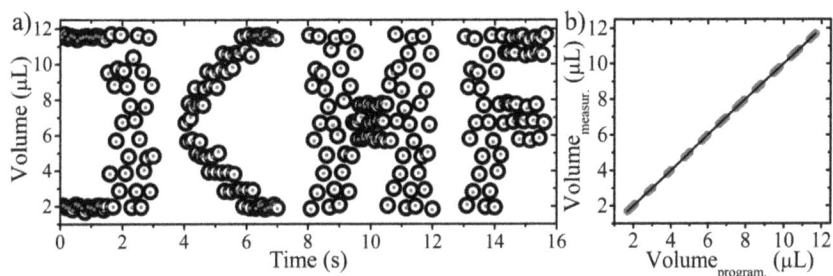

3.6. Durability of the Valve

We demonstrated the use of this system to issue droplets of varied, pre-defined, volumes at a constant frequency of forming. This scheme of operation may be particularly suitable for screening applications. To build robust systems on the basis of the design presented here one should consider using a proper material for the membrane. We checked that the foils that we used typically survived *ca.* 5 million cycles of opening and closing the valve and then quickly degraded.

Higher durability of the membrane would be also advisable as our system is external to the chip, and therefore can be used many times in different setups. This feature dramatically lowers the complication of disposable parts (chips) of the system and therefore the cost of a single measurement. This feature is particularly important for current applications of the micro-droplet techniques.

4. Conclusions

We reported an automated system for on-demand generation of micro-droplets, ranging in volume from 1 nL to 20 μL at frequencies ranging up to 400 Hz. The system uses a piezoelectric valve and offers high precision in administering of volume of liquid: we measured the standard deviation of the volumes of droplets to be well below 1% of the mean for the given set of values of the control parameters (Table S1 in supplementary).

The experiments prove that the performance of our system is independent from the cross section of the channel in the tested regime (100×100–400×400 μm^2) and the applied membrane features sufficient durability. Generated droplets have volumes of pre-programmed values, and their appearance in the microchannel is strictly defined. Similar behavior of our piezoelectric valve is expectable in microchannels with width lower than 100 μm and the droplets should be generated with a good accuracy even at pL volume range, this however requires that the pressure head in the resistive capillary connecting the valve with the chip be at least 100 times greater than the typical pressure head in the microfluidic chip.

The presented system requires pre-filtered fluids to avoid blocking of the valve. This feature prevents also obstruction of the microchannels with debris, which is highly important especially in biological and medical applications.

The advantage of the presented system over the systems based on electromagnetic valves is the much higher frequency of droplet-on-demand generation with the same, or even higher precision in administering the volumes. The system that we report here can be integrated into screening systems operating at tens of Hz and on truly nano-liter volumes of the samples—a combination that surpasses the titier plate robotics by more than two orders of magnitude in the speed of operation and by at least an order of magnitude in the reaction volume.

Supplementary Materials

Supplementary materials can be accessed at: http://www.mdpi.com/2072-666X/5/4/1002/s1.

Acknowledgments

Project operated within the European Research Council Starting Grant 279647. Slawomir Jakiela and Pawel R. Debski also acknowledges financial support from the Polish Ministry of Science and Higher Education under the grant Iuventus Plus IP2012 015172.

Author Contributions

Slawomir Jakiela, Pawel R. Debski and Piotr Garstecki designed the experiments. Slawomir Jakiela and Bogdan Dabrowski developed and assembled the piezolectric valve. Slawomir Jakiela, Pawel R. Debski executed the experiments and together with Piotr Garstecki analyzed the results and wrote the manuscript.

Conflicts of Interest

The authors declare there are no conflicts of interest.

References

1. Theberge, A.B.; Courtois, F.; Schaerli, Y.; Fischlechner, M.; Abell, C.; Hollfelder, F.; Huck, W.T.S. Microdroplets in microfluidics: An evolving platform for discoveries in chemistry and biology. *Angew. Chem. Int. Ed. Engl.* **2010**, *49*, 5846–5868.

2. Pompano, R.R.; Liu, W.; Du, W.; Ismagilov, R.F. Microfluidics using spatially defined arrays of droplets in one, two, and three dimensions. *Annu. Rev. Anal. Chem.* **2011**, *4*, 59–81.

3. Squires, T.M.; Quake, S.R. Microfluidics: Fluid physics at the nanoliter scale. *Rev. Mod. Phys.* **2005**, *77*, 977–1026.

4. Leng, X.; Zhang, W.; Wang, C.; Cui, L.; Yang, C.J. Agarose droplet microfluidics for highly parallel and efficient single molecule emulsion PCR. *Lab Chip* **2010**, *10*, 2841–2843.

5. Wang, M.S.; Nitin, N. Rapid detection of bacteriophages in starter culture using water-in-oil-in-water emulsion microdroplets. *Appl. Microbiol. Biotechnol.* **2014**, *98*, 8347–8355.

6. Witters, D.; Sun, B.; Begolo, S.; Rodriquez-Manzano, J.; Robles, W.; Ismagilov, R.F. Digital biology and chemistry. *Lab Chip* **2014**, *14*, 3225–3232.

7. Teste, B.; Ali-Cherif, A.; Viovy, J.L.; Malaquin, L. A low cost and high throughput magnetic bead-based immuno-agglutination assay in confined droplets. *Lab Chip* **2013**, *13*, 2344–2349.

8. Churski, K.; Korczyk, P.; Garstecki, P. High-throughput automated droplet microfluidic system for screening of reaction conditions. *Lab Chip* **2010**, *10*, 816–818.

9. Churski, K.; Kaminski, T.S.; Jakiela, S.; Kamysz, W.; Baranska-Rybak, W.; Weibel, D.B.; Garstecki, P. Rapid screening of antibiotic toxicity in an automated microdroplet system. *Lab Chip* **2012**, *12*, 1629–1637.

10. Perccin, G.; Levin, L.; Khuri-Yakub, B.T. Piezoelectrically actuated droplet ejector. *Rev. Sci. Instrum.* **1997**, *68*, 4561–4563.

11. Chen, C.H.; Cho, S.H.; Tsai, F.; Erten, A.; Lo, Y.-H. Microfluidic cell sorter with integrated piezoelectric actuator. *Biomed. Microdevices* **2009**, *11*, 1223–1231.

12. Bransky, A.; Korin, N.; Khoury, M.; Levenberg, S. A microfluidic droplet generator based on a piezoelectric actuator. *Lab Chip* **2009**, *9*, 516–520.

13. Shemesh, J.; Bransky, A.; Khoury, M.; Levenberg, S. Advanced microfluidic droplet manipulation based on piezoelectric actuation. *Biomed. Microdevices* **2010**, *12*, 907–914.

14. Shemesh, J.; Nir, A.; Bransky, A.; Levenberg, S. Coalescence-assisted generation of single nanoliter droplets with predefined composition. *Lab Chip* **2011**, *11*, 3225–3230.

15. Xu, J.; Attinger, D. Drop on demand in a microfluidic chip. *J. Micromech. Microeng.* **2008**, *18*, 065020.

16. Abate, A.R.; Hung, T.; Mary, P.; Agresti, J.J.; Weitz, D.A. High-throughput injection with microfluidics using picoinjectors. *Proc. Natl. Acad. Sci.* **2010**, *107*, 19163–19166.

17. Tang, J.; Jofre, A.M.; Kishore, R.B.; Reiner, J.E.; Greene, M.E.; Lowman, G.M.; Denker, J.S.; Willis, C.C.C.; Helmerson, K.; Goldner, L.S. Generation and mixing of subfemtoliter aqueous droplets on demand. *Anal. Chem.* **2009**, *81*, 8041–8047.

18. Thorsen, T.; Roberts, R.W.; Arnold, F.H.; Quake, S.R. Dynamic pattern formation in a vesicle-generating microfluidic device. *Phys. Rev. Lett.* **2001**, *86*, 4163–4166.

19. Garstecki, P.; Fuerstman, M.J.; Stone, H.A.; Whitesides, G.M. Formation of droplets and bubbles in a microfluidic T-junction—Scaling and mechanism of break-up. *Lab Chip* **2006**, *6*, 437–446.

20. Jakiela, S.; Zaslona, J.; Michalski, J.A. Square wave driver for piezoceramic actuators. *Actuators* **2012**, *1*, 12–20.

21. Churski, K.; Nowacki, M.; Korczyk, P.M.; Garstecki, P. Simple modular systems for generation of droplets on demand. *Lab Chip* **2013**, *13*, 3689–3697.

22. Jakiela, S.; Kaminski, T.S.; Cybulski, O.; Weibel, D.B.; Garstecki, P. Bacterial growth and adaptation in microdroplet chemostats. *Angew. Chem. Int. Ed. Engl.* **2013**, *52*, 8908–8911.

23. Cao, J.; Kürsten, D.; Schneider, S.; Knauer, A.; Günther, P.M.; Köhler, J.M. Uncovering toxicological complexity by multi-dimensional screenings in microsegmented flow: Modulation of antibiotic interference by nanoparticles. *Lab Chip* **2012**, *12*, 474–484.

5

Microfluidic Vortex Enhancement for on-Chip Sample Preparation

Anna Haller [1,*], Andreas Spittler [2], Lukas Brandhoff [3], Helene Zirath [4],
Dietmar Puchberger-Enengl [1], Franz Keplinger [1] and Michael J. Vellekoop [3]

[1] Institute of Sensor and Actuator Systems, Vienna University of Technology, Gusshausstrasse 27-29/E366, 1040 Vienna, Austria; E-Mails: dietmar.puchberger-enengl@tuwien.ac.at (D.P.-E.); franz.keplinger@tuwien.ac.at (F.K.)

[2] Core Facility Flow Cytometry & Department of Surgery, Research Laboratories, Center of Translational Research, Medical University of Vienna, Lazarettgasse 14, 1090 Vienna, Austria; E-Mail: andreas.spittler@meduniwien.ac.at

[3] Institute of Microsensors, -Actuators and -Systems (IMSAS) & Microsystems Center Bremen (MCB), University of Bremen, Otto-Hahn-Allee 1, 28359 Bremen, Germany; E-Mails: lbrandhoff@imsas.uni-bremen.de (L.B.); mvellekoop@imsas.uni-bremen.de (M.J.V.)

[4] Health and Environment Department, Austrian Institute of Technology, Muthgasse 11, 1190 Vienna, Austria; E-Mail: helene.zirath.fl@ait.ac.at

* Author to whom correspondence should be addressed; E-Mail: anna.haller@tuwien.ac.at

Academic Editor: Joost Lötters

Abstract: In the past decade a large amount of analysis techniques have been scaled down to the microfluidic level. However, in many cases the necessary sample preparation, such as separation, mixing and concentration, remains to be performed off-chip. This represents a major hurdle for the introduction of miniaturized sample-in/answer-out systems, preventing the exploitation of microfluidic's potential for small, rapid and accurate diagnostic products. New flow engineering methods are required to address this hitherto insufficiently studied aspect. One microfluidic tool that can be used to miniaturize and integrate sample preparation procedures are microvortices. They have been successfully applied as microcentrifuges, mixers, particle separators, to name but a few. In this work, we utilize a novel corner structure at a sudden channel expansion of a microfluidic chip to enhance the formation of a microvortex. For a maximum area of the microvortex, both chip geometry and corner structure were optimized with a computational fluid dynamic (CFD) model. Fluorescent particle trace measurements with the optimized design prove that the corner structure increases the size of the vortex. Furthermore, vortices are induced by the corner structure at low flow rates while no recirculation is observed without a corner structure. Finally, successful separation of plasma from human blood was accomplished, demonstrating a potential application for clinical sample preparation. The extracted plasma was characterized by a flow cytometer and compared to plasma obtained from a standard benchtop centrifuge and from chips without a corner structure.

Keywords: microfluidic sample preparation; microvortex enhancement; on-chip human blood plasma separation

1. Introduction

A primary objective in microfluidics is the miniaturization of laboratory devices and subsystems to so-called lab-on-a-chip (LoC) systems. In this field, many publications focus on the microfluidic investigation of the properties and the handling of various analytes. However, a crucial precondition to obtain a fully integrated LoC is sample preparation directly on the chip. The preparation steps include focusing, separating and concentrating cells or particles in suspension and mixing and exposing targets to samples, among others. While numerous analysis techniques have been developed for on-chip application, the necessary preparation steps are often performed off-chip [1]. Consequently, new flow engineering methods have to be developed and applied to meet the challenges faced prior to the actual analysis [2,3].

Currently, considerable efforts towards the development of microfluidic chips based on inertial fluid forces can be observed [4]. Recirculation areas, so called microvortices, prove to be a versatile and powerful tool in microfluidics. Shelby *et al.* were among the first to investigate the high radial acceleration present in microvortices and their effect on particles and cells [5–7].

The operation principle of several microfluidic devices is based on the creation of microvortices. They are utilized for on-chip microcentrifuges [8–10], to focus or separate particles [11–13], to manipulate cells [7,14], to trap particles [15,16], to enrich rare cells, e.g., circulating tumor cells [17,18], to fabricate micromixers [19–21], to synthesize size-controlled nanoparticles [22], or to extract plasma from blood [23,24]. All these works rely on the efficient generation of microvortices.

A thorough review article on microfluidic separation and sorting, covering the application of microvortices, was recently published [25]. Numerical investigation of microvortices or recirculation areas has been performed under diverse settings [14,26–29]. Apart from the fact that most continuous flow systems rely on a pumping mechanism, microvortex generating systems can be divided into passive, *i.e.*, utilizing hydrodynamic effects, and active ones. Microvortices are actively formed by an oscillating microbubble [30], by nickel nanowires in rotating magnetic fields [16], by piezoelectric actuation [31] or by an ionic wind mechanism [32–34]. Active mechanisms to generate microvortices increase complexity of fabrication and operation and, therefore, susceptibility to failures.

Passive microvortices are created by channel geometries like sudden expansions in microscale channels [7,12,15,18,19], cylindrical microcavities [35], trapezoidal side chambers [14], microbifurcations [28], embedded obstacles inside the flow channel [21,36,37], two counterflowing liquid streams [9] or at the boundary layer between sheath and center stream [22].

Microfluidic channels with sudden expansions like backward-facing steps are a well-established structure to induce microvortices [38]. Laminar flow characteristics demand high flow rates to generate vortices at the microscale. In case of rare and costly samples, e.g., neonatal blood or bone marrow fluid, elevated flow rates are challenging.

We introduce a geometry that enhances the vortex growth, thereby reducing flow rate and sample consumption. It consists of a corner structure that is located directly before the expansion channel and protrudes into the channel under a defined angle (Figure 1a). This design allows reducing the volume flow rate significantly while maintaining vortex area and vortex shape. First, the geometric parameters of the channel and the flow rate were systematically varied in a simulation study to achieve an optimal design for a maximum vortex area. Second, the chip was fabricated with silicon technology and characterized by a fluorescence streamline visualization experiment. Finally, as proof-of-concept, the novel channel design was successfully applied as a human blood plasma extractor.

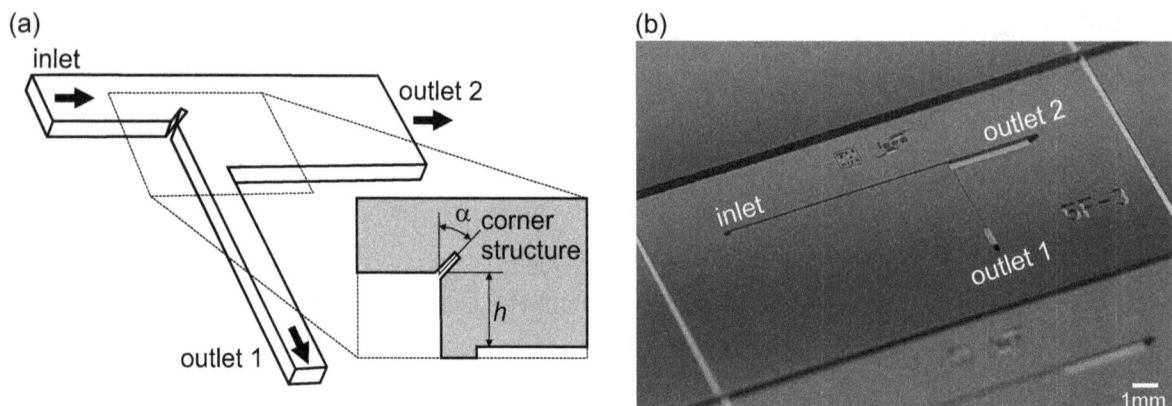

Figure 1. (**a**) Schematic of the microfluidic chip showing the backward-facing step and the novel corner structure in detail (angle of corner structure α, step height h); (**b**) Photograph of a fabricated chip.

2. Materials and Methods

In contrast with designs with plain backward-facing steps, the channel in our approach is initially constricted by a corner structure before it features a sudden expansion. This corner structure changes the flow profile in the region of the channel expansion, thereby enhancing the area of microvortices. The complete chip design depicted in Figure 1a comprises one inlet, where the sample enters the microfluidic channel, two outlets and the corner structure, 40 μm long and 10 μm wide at the edge of the expansion [39].

This influence is analyzed by a simulation study in Section 3.1 and measured using fluorescent particles as tracers in Section 3.2. To quantify the effect of the corner structure, the area of the microvortex is determined. We optimized our design for a maximum vortex area. As shown by

Chiu *et al.* [7] particles with higher density than the buffer solution are ejected from the microvortices. Hence, a larger vortex can increase the separation efficiency. The vortex area can directly be assessed in simulations as well as experiments and, hence, allows a comparison of CFD analysis and microfluidic measurements. To find the maximum value of the vortex area an optimization was performed by a series of simulations with varying angles of the corner structure α and step heights h. By varying the angle α the effective protrusion length of the corner structure into the channel is considered. To prevent clogging and limit the shear force in the narrowed area, we choose to fix the maximum length of the corner structure to 40% of the channel width.

The channel widths of the inlet and outlet 1 are 100 μm and 50 μm, respectively. The width of outlet 2 is the sum of the step height h and the width of the inlet channel. A uniform channel depth of 100 μm was chosen for the device.

2.1. Numerical Models

The laminar flow profile was calculated by applying the incompressible Navier-Stokes model on COMSOL Multiphysics 4 platform. The numerical model was simplified by assuming a steady state flow regime. At the channel wall a non-slip boundary condition was applied. The physical properties of water were applied to the fluid in the model (dynamic viscosity $\mu = 10^{-3}$ Pa $\cdot$ s, density $\rho = 1000$ kg/m^3).

For the validation of the streamline visualization experiment, the liquid phase was modeled as a solution of 14% *w/w* sucrose in water (dynamic viscosity $\mu_{14\%\text{sucrose}} = 1.53$ m Pa $\cdot$ s, density $\rho_{14\%\text{sucrose}} = 1056$ kg/m^3) utilizing a three-dimensional model of the channel structure. For the inlet a volume flow rate was set, whereas for the outlets a hydrodynamic resistance equivalent to the fluidic setup was applied.

2.2. Device Fabrication

The microfluidic chip was fabricated applying standard micromachining techniques with a silicon (Si) wafer (350 μm thick, <100> orientation, $\phi = 100$ mm, double side polished) as substrate and a borosilicate glass wafer (500 μm thick, $\phi = 100$ mm, double side polished) as cover. Initially, inlet and outlets for the fluid connections were patterned on the bottom side of the Si wafer by lithographic steps. The microfluidic channel was fabricated into the top side of the Si wafer. On both sides Si was structured using deep reactive ion etching (DRIE). Finally, the glass wafer was anodically bonded onto the top side of the silicon wafer, sealing the microfluidic channel structure. A detailed description of the fabrication of microfluidic devices using glass and silicon has been published recently [40]. On a 100 mm wafer 20 devices are fabricated. Figure 1b shows a photograph of a diced microfluidic chip.

2.3. Particle Suspension and Blood Preparation

For the streamline visualization measurement in Section 3.2 a sample with 0.1% *v/v* fluorescent green 3 μm sized polystyrene particles (G0300, Distrilab B.V., Leusden, The Netherlands) was prepared by diluting the suspension with a density-matched solution of 14% *w/w* sucrose in deionized (DI) water. To avoid sticking of the fluorescent particles to the channel walls, the chip was preflushed with a sterile-filtered solution of 2 g/L alginic acid sodium salt (Sigma-Aldrich, St. Louis, MO, USA) in DI water. After a settling time of 2 min the visualization of the streamlines was started.

For the plasma separation experiments blood was collected by venipuncture from healthy donors taking part in a clinical study, approved by the local institutional ethics board. The blood sample was diluted with a phosphate buffered saline (PBS) solution by a factor of 20 which led to the best performance for the current design. Similar dilution rates are reported by other groups using cell deviation mechanisms [41]. Generally, there is a trade-off between dilution rate and separation efficiency. To prevent sedimentation of blood cells the sample was continuously agitated with a magnetic syringe stirrer (neMIX, Cetoni, Korbussen, Germany) before entering the chip.

2.4. Experimental Setup

Fluorescence images were captured with a Nikon AZ 100 upright microscope (Nikon, Tokyo, Japan), equipped with an AZ Plan Fluor 5x objective and a digital camera (D5100, Nikon). The exposure time was varied from 1/200 s to 3 s according to the light intensity and the applied flow rate. The fluorescent particles were stimulated with a mercury arc lamp (Intensilight C-HGFI, Nikon) and an optical single-band filter set (GFP-3035D-NTE, Semrock, Rochester, NY, USA).

The microfluidic device was connected with standard PEEK (polyether ether ketone) tubings (outer diameter $\phi = 794\,\mu m$, inner diameter $\phi = 250\,\mu m$) via thermoplastic bonded-port connectors (Labsmith CapTite, Mengel Engineering, Virum, Denmark), which are glued onto the backside of the silicon device, in combination with a screwed-in fitting (Labsmith CapTite). This world-to-chip interface is illustrated in the electronic supplementary information, Figure S1. To provide stable flow rates a syringe pump (neMESYS, Cetoni, Korbussen, Germany) was used in all experiments. For the blood plasma separation experiments another syringe pump (Cetoni neMESYS) was used to set the plasma outflow rate.

2.5. Procedures for Image Analysis

To acquire the size of the vortex area, all images were analyzed with Photoshop CS5 (Adobe Systems Incorporated, San Jose, CA, USA). In the simulated streamline plots the perimeter of the outmost circulating streamline was traced. The area of the recirculation was measured using the histogram tool.

Micrographs of the fluorescent particle traces were initially posterized. This simplifies the differentiation between circulating and passing-by streamlines. Subsequent steps are equal to the postprocessing of the simulation results.

3. Measurements and Results

An optimization of the corner structured channel design was performed utilizing numerical models in Section 3.1. To determine the microvortex enhancement and compare the simulations with the measurements, a streamline visualization using fluorescent particles was conducted in Section 3.2. A potential application of sudden expansion channels for on-chip sample preparation is the separation of human blood plasma. Inducing microvortices in expansion channels increases the plasma extraction yield [24]. In Section 3.3 the effect of enhancing the vortex area by the corner structured chip is investigated in respect to separation efficiency.

3.1. Design Optimization

In order to optimize the channel geometry and the corner structure for a maximum vortex area, a computational analysis was performed according to Section 2.1. The computational data in Figure 2 shows a representative velocity and streamline profile of a backward-facing step in the channel with (a) and without (b) corner structure at a volume flow rate of 150 µL/min. This comparison illustrates the formation of microvortices by the corner structure whereby no recirculation can be observed in designs with a plain backward-facing step.

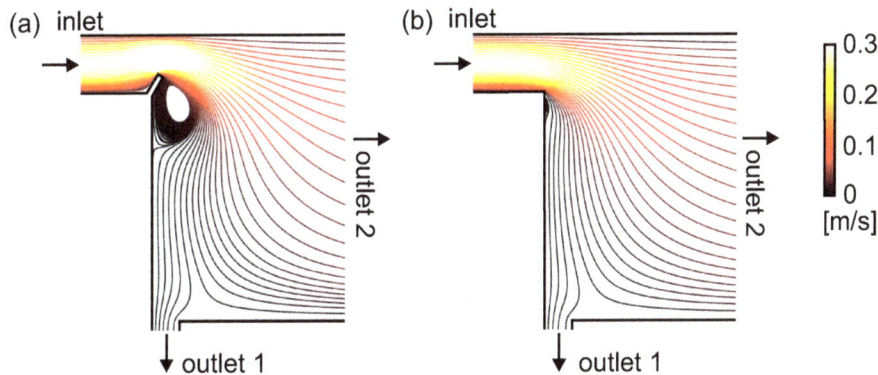

Figure 2. Velocity streamlines in a sudden expansion channel ($h = 400\,\mu m$) with (**a**) and without (**b**) the corner structure (width $\times$ length: $10\,\mu m \times 40\,\mu m$, $\alpha = -30°$) at $Re = 25$. Re was calculated at the inlet channel (width $\times$ depth: $100\,\mu m \times 100\,\mu m$) with inlet flow rate 150 µL/min. The color of streamlines indicates the velocity magnitude (m/s).

The Reynolds number Re is the ratio of inertial to viscous forces in flow ($Re = \rho u D_h / \mu$, with ρ, u, D_h and μ being the density, mean downstream velocity, hydraulic diameter and dynamic viscosity, respectively). The hydraulic diameter D_h can be defined as $D_h = 4 \cdot A/P$, with A being the cross sectional area and P the wetted perimeter of the microfluidic channel prior to the expansion. For the proposed chip design and flow rates, Re is in the range of 5 to 60 in the inlet channel, stating the laminar flow regime.

In the CFD analysis, three parameters were varied to maximize the vortex area: step height h, angle of the corner structure α and volume flow rate. To visualize the microvortices in the microfluidic channel, velocity streamlines were plotted. Next, the microvortex area was calculated using image processing tools as described in Section 2.5. Figure 3 summarizes the results of the simulation study. The three multiple line graphs (a), (b) and (c) represent the volume flow rates 100 µL/min, 200 µL/min and 300 µL/min, respectively. The microvortex area is plotted against the angle of the corner structure α, including the case of no corner structure, *i.e.*, a channel with a plain backward-facing step design (values at the ordinate). An angle of $0°$ corresponds to a vertical corner structure protruding into the channel, *i.e.*, perpendicular to the flow direction. Each line in the plot corresponds to an individual channel step height h.

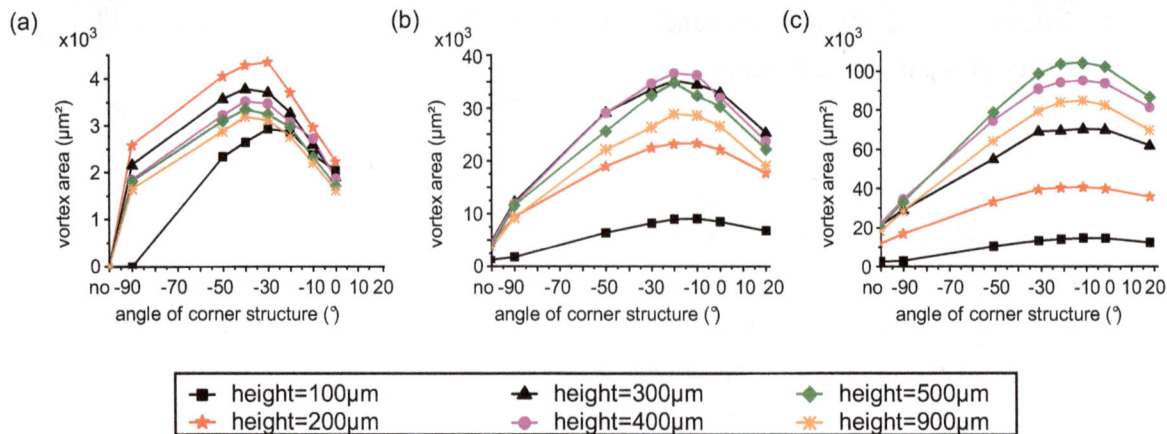

Figure 3. Microvortex area as a function of the angle of the corner structure calculated in the simulation study for different step heights and inlet volume flow rates of 100 μL/min (**a**) 200 μL/min (**b**) and 300 μL/min (**c**).

Independently of the angle α and the step height h, the corner structure generally increases the microvortex area.

To identify the design which is best applicable for the wide flow rate range, the vortex area was normalized. Normalization enables a comparison of microvortex areas at different flow velocities with each other. Hence, three values, corresponding to 100 μL/min, 200 μL/min and 300 μL/min, represent the performance of a channel design. Subsequently, the mean value $\overline{x}$ and the standard deviation s_x for each of the different channel geometries were calculated. Based on this data, the geometric dimension was chosen with respect to the highest mean value while having a low standard deviation ($s_x < 10\%$). This optimization scheme led to the optimum design featuring a step height h of 400 μm and an angle α of $-30°$ ($\overline{x} = 87\%$, $s_x = 6\%$).

In summary, by applying the optimized corner structure to a sudden expansion channel (inlet channel width 100 μm, step height 400 μm, flow rate 200 μL/min) the microvortex area is increased by more than a factor of 4. This optimized design creates a channel constriction just before the expansion, resulting in a minimum channel width of about 65 μm. A plain backward-facing step design with an equivalent narrowed inlet channel width of 65 μm causes a smaller vortex. The improvement in microvortex area of the corner structured compared to a uniform narrowed channel for a volume flow rate of 200 μL/min amounts to approximately 40%.

3.2. Streamline Visualization

To visualize and characterize the enlargement of the microvortex area by the corner structure a constant flow of a suspension with fluorescent particles was injected to trace the streamlines. Figure 4 compares the fluorescent particle trajectories of chips with (a) and without (b) corner structure at a constant flow rate of 150 μL/min. These measurements verify the previous simulation study, depicted in Figure 2. Introducing the corner structure lowers the volume flow rate necessary for vortex formation onset. Additional micrographs of fluorescent particle streamline measurements, comparing the plain backward-facing step channel *versus* the optimized corner structured chip, are available as electronic supplementary material in Figure S2.

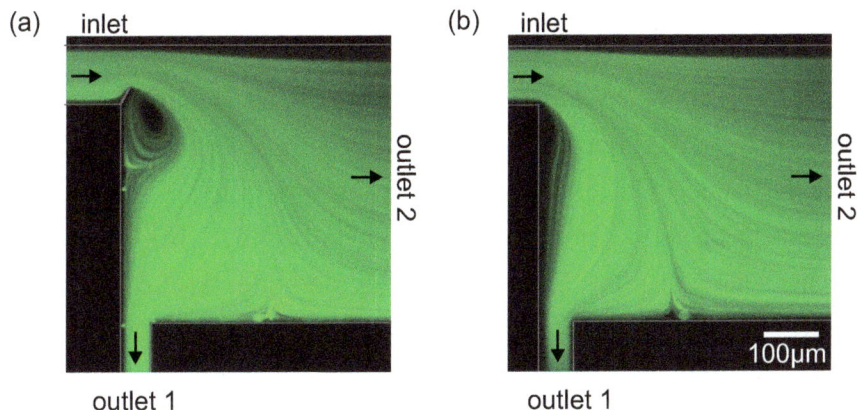

Figure 4. Micrographs of fluorescent particle streamlines in the optimized sudden expansion channel with (**a**) and without (**b**) corner structure at $Re = 25$.

To characterize the formation and the enhancement of the microvortices the solution with the fluorescent particles was pumped through the microfluidic device at volume flow rates, ranging from 50 to 350 µL/min. Figure 5 compares the vortex areas of the simulation study (blue circles, black triangles) with the measured values (green rectangles, red diamonds). These results validate the simulation, which was performed according to Section 2.1. The difference between simulated and measured data arises from the fact that simulated streamlines are compared with measured particle trajectories and the neglection of particle-particle interactions.

The greatest influence on vortex formation and enhancement by the corner structure is observed at low flow rates. At a volume flow rate of 200 µL/min the vortex area is enhanced by a factor of 4.4, whereas at higher flow rates (up to 350 µL/min) the factor is still greater than 2.5. At flow rates below 150 µL/min the formation of a microvortex is only provided by the channel with the corner structure.

Furthermore, this figure reveals that, by applying the corner structure, the flow rate can be reduced without a decrease in vortex size.

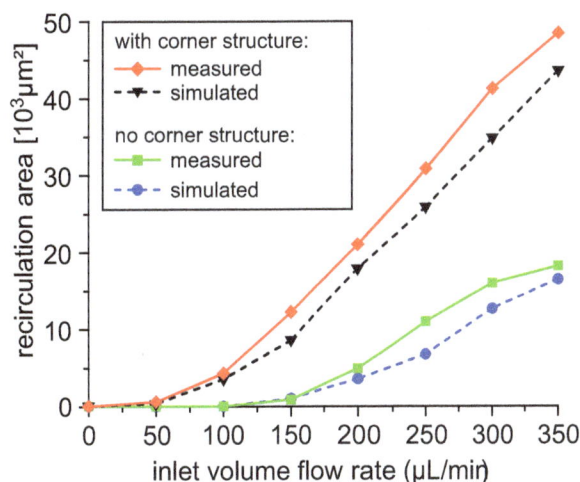

Figure 5. Comparison of the simulated (dashed lines) and measured (solid lines) vortex area in the sudden expansion channel with and without corner structure as a function of the inlet volume flow rate.

3.3. Human Blood Plasma Separation

As an example of integrated sample preparation, the corner structured chip was tested for the ability to separate plasma from anticoagulated human blood. Figure 6a depicts a micrograph of the separation using a sudden expansion channel with corner structure at a flow rate of 200 µL/min. During the measurements no hemolysis has been observed. As the red blood cells (RBCs), which constitute about 95% of all blood cells, are denser than the liquid phase, they do not enter the vortex. Hence, a larger vortex means a larger cell-free area. In addition, the corner structure as well as the sudden expansion geometrically enhances the cell-free layer, which forms due to hydrodynamic lift of deformable cells in shear flow. This phenomenon has been systematically investigated by Faivre *et al.* [42]. Thus, RBCs are concentrated at outlet 2, whereas plasma is extracted at outlet 1.

(a)   (b)

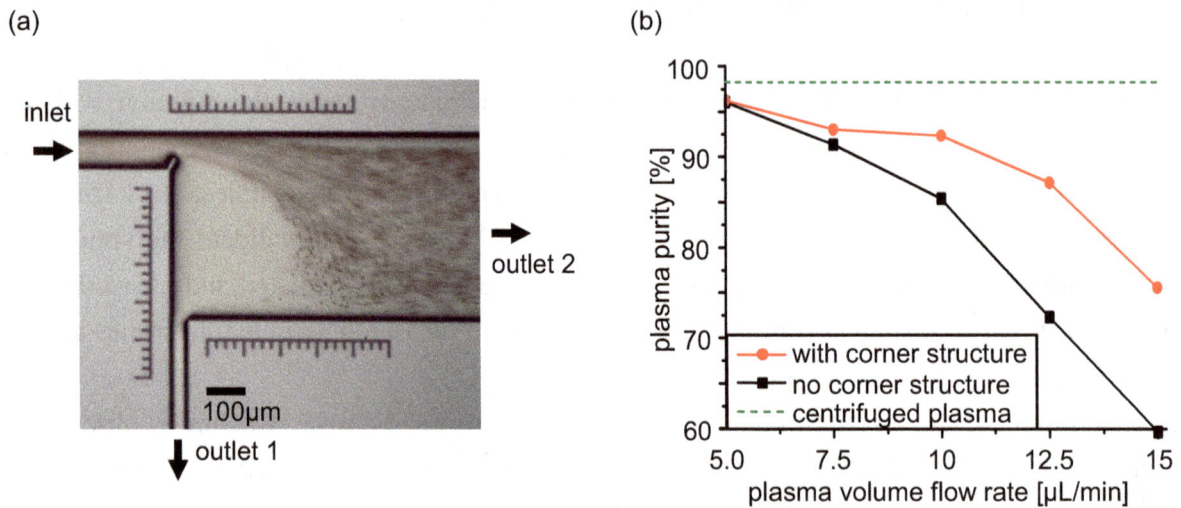

Figure 6. (a) Bright-field image of human blood plasma separation experiment at a flow rate of 200 µL/min. The concentrated blood stream exits at outlet 2. (b) Purity of plasma generated by the microfluidic chip with and without corner structure in comparison to centrifuged plasma.

A flow cytometer (Gallios, Beckman Coulter Inc., Miami, FL, USA) is used to determine the purity (the percentage of removed cells) and the quality of the generated plasma. The purity p is defined as $p = 1 - c_p/c_f$, where c_p is the number of cells in the plasma fraction and c_f is the number of cells in the feed (inlet) fraction.

Figure 6b compares the purity of conventional, *i.e.*, generated with a bench-top centrifuge, plasma (dotted green) with plasma extracted from the microfluidic chip with (red circles) and without (black rectangles) corner structure. The inlet volume flow rate was kept constant at 200 µL/min, the plasma extraction rate was varied between 5 µL/min and 15 µL/min, corresponding to a yield (the percentage of plasma over the total volume) up to 7.5%.

At low plasma flow rates the designs with and without corner structure perform equally, attaining a plasma purity of over 95%. The purity of centrifuged plasma was about 98% due to the presence of microparticles <3 µm. In the conducted flow cytometer measurement this is the maximum purity reached with standard benchtop centrifuges. By applying the corner structure at elevated plasma flow rates, a higher purity is reached compared to designs with a plain backward-facing step. At a flow rate of 12.5 μL/min the particle count in the plasma sample is reduced by more than 50% compared to the design without corner structure. This corresponds to an increased purity of 87%. As seen, especially at higher flow rates the corner structure increases plasma purity, reducing the trade-off between throughput and quality.

This proof-of-concept confirms that the corner structure increases the separation efficiency by enhancing the vortex area. Microscale blood fractionation is a research area of great interest. A variety of microfluidic devices with the aim of integrating blood plasma extraction have been investigated and reviewed [41,43,44], thereby harnessing the advantages of miniaturized systems. Although, in comparison the yield of the presented system is lower, there are several possibilities to improve the performance, e.g., with a symmetric geometry [45], parallelization [46] or installing units in series [23,47].

4. Conclusions

This work reports on a microfluidic chip which utilizes a novel corner structure in a sudden expansion channel to generate and enhance microvortices under laminar flow conditions. By optimizing the geometric properties, the vortex area is increased by more than a factor of 4 compared to plain backward-facing steps at 200 μL/min, which was proven by fluorescent streamline visualization. In addition to the enlarged microvortex area, the new design exhibits a lower minimal volume flow rate that triggers microvortex formation.

Consequently, the proposed channel design can be used to improve separation of cells and particles in suspension, hereby meeting typical challenges of on-chip sample preparation. The extraction of human blood plasma is presented as an application of the proposed chip design. An enhanced purity of more than 90% is reached with a cell count reduction by a factor of 2 compared to channels without corner structure.

Channel protrusions, such as the presented corner structure, represent a simple method to improve the performance of microvortex-based microfluidics.

Supplementary Materials

Supplementary materials can be accessed at: http://www.mdpi.com/2072-666X/6/2/239/s1.

Acknowledgments

For the sensor fabrication and technical support we thank Edda Svasek and Peter Svasek (MEMS Technology Lab of the Institute of Sensor and Actuator Systems and Center for Micro- and Nanostructures, Vienna University of Technology). We gratefully acknowledge the support of the Austrian Research Promotion Agency, grant No. 829651.

Author Contributions

Anna Haller conceived the idea for the project, designed and performed the experiments, analyzed the data and wrote the paper. Andreas Spittler designed and performed the flow cytometer experiments and analyzed the data. Lukas Brandhoff performed experiments and revised the manuscript. Helene Zirath performed experiments and revised the manuscript. Dietmar Puchberger-Enengl contributed reagents and materials and revised the manuscript. Franz Keplinger revised the manuscript and provided senior expertise. Michael J. Vellekoop conceived the idea for the project, revised the manuscript and provided senior expertise.

Conflicts of Interest

The authors declare no conflict of interest.

References

1. Mach, A.J.; Adeyiga, O.B.; di Carlo, D. Microfluidic sample preparation for diagnostic cytopathology. *Lab Chip* **2013**, *13*, 1011–1026.
2. Culbertson, C.T.; Mickleburgh, T.G.; Stewart-James, S.A.; Sellens, K.A.; Pressnall, M. Micro total analysis systems: Fundamental advances and biological applications. *Anal. Chem.* **2014**, *86*, 95–118.
3. Verpoorte, E. Microfluidic chips for clinical and forensic analysis. *Electrophoresis* **2002**, *23*, 677–712.
4. Amini, H.; Lee, W.; di Carlo, D. Inertial microfluidic physics. *Lab Chip* **2014**, *14*, 2739–2761.
5. Shelby, J.P.; Lim, D.S.W.; Kuo, J.S.; Chiu, D.T. Microfluidic systems: High radial acceleration in microvortices. *Nature* **2003**, *425*, 38.
6. Shelby, J.P.; Chiu, D.T. Controlled rotation of biological micro- and nano-particles in microvortices. *Lab Chip* **2004**, *4*, 168–170.
7. Chiu, D.T. Cellular manipulations in microvortices. *Anal. Bioanal. Chem.* **2007**, *387*, 17–20.
8. Mach, A.J.; Kim, J.H.; Arshi, A.; Hur, S.C.; di Carlo, D. Automated cellular sample preparation using a Centrifuge-on-a-Chip. *Lab Chip* **2011**, *11*, 2827–2834.
9. Pertaya-Braun, N.; Baier, T.; Hardt, S. Microfluidic centrifuge based on a counterflow configuration. *Microfluid. Nanofluid.* **2011**, *12*, 317–324.
10. Lee, J.; Ha, J.; Bahk, Y.; Yoon, S.; Arakawa, T.; Ko, J.; Shin, B.; Shoji, S.; Go, J. Microfluidic centrifuge of nano-particles using rotating flow in a microchamber. *Sens. Actuators B Chem.* **2008**, *132*, 525–530.
11. Park, J.S.; Song, S.H.; Jung, H.I. Continuous focusing of microparticles using inertial lift force and vorticity via multi-orifice microfluidic channels. *Lab Chip* **2009**, *9*, 939–948.
12. Wang, X.; Zhou, J.; Papautsky, I. Vortex-aided inertial microfluidic device for continuous particle separation with high size-selectivity, efficiency, and purity. *Biomicrofluidics* **2013**, *7*, 044119.
13. Hsu, C.H.; Di Carlo, D.; Chen, C.; Irimia, D.; Toner, M. Microvortex for focusing, guiding and sorting of particles. *Lab Chip* **2008**, *8*, 2128–2134.
14. Zhang, W.; Frakes, D.H.; Babiker, H.; Chao, S.H.; Youngbull, C.; Johnson, R.H.; Meldrum, D.R. Simulation and experimental characterization of microscopically accessible hydrodynamic microvortices. *Micromachines* **2012**, *3*, 529–541.
15. Zhou, J.; Kasper, S.; Papautsky, I. Enhanced size-dependent trapping of particles using microvortices. *Microfluid. Nanofluid.* **2013**, *15*, 611–623.

16. Petit, T.; Zhang, L.; Peyer, K.E.; Kratochvil, B.E.; Nelson, B.J. Selective trapping and manipulation of microscale objects using mobile microvortices. *Nano Lett.* **2012**, *12*, 156–160.

17. Hur, S.C.; Mach, A.J.; di Carlo, D. High-throughput size-based rare cell enrichment using microscale vortices. *Biomicrofluidics* **2011**, *5*, 022206.

18. Sollier, E.; Go, D.E.; Che, J.; Gossett, D.R.; O'Byrne, S.; Weaver, W.M.; Kummer, N.; Rettig, M.; Goldman, J.; Nickols, N.; *et al.* Size-selective collection of circulating tumor cells using Vortex technology. *Lab Chip* **2014**, *14*, 63–77.

19. Lee, M.G.; Choi, S.; Park, J.K. Rapid multivortex mixing in an alternately formed contraction-expansion array microchannel. *Biomed. Microdevices* **2010**, *12*, 1019–1026.

20. Lee, J.; Kwon, S. Mixing efficiency of a multilamination micromixer with consecutive recirculation zones. *Chem. Eng. Sci.* **2009**, *64*, 1223–1231.

21. Shih, T.R.; Chung, C.K. A high-efficiency planar micromixer with convection and diffusion mixing over a wide Reynolds number range. *Microfluid. Nanofluid.* **2007**, *5*, 175–183.

22. Kim, Y.; Lee Chung, B.; Ma, M.; Mulder, W.J.M.; Fayad, Z.A.; Farokhzad, O.C.; Langer, R. Mass production and size control of lipid-polymer hybrid nanoparticles through controlled microvortices. *Nano Lett.* **2012**, *12*, 3587–3591.

23. Marchalot, J.; Fouillet, Y.; Achard, J.L. Multi-step microfluidic system for blood plasma separation: Architecture and separation efficiency. *Microfluid. Nanofluid.* **2014**, *17*, 167–180.

24. Sollier, E.; Cubizolles, M.; Fouillet, Y.; Achard, J.L. Fast and continuous plasma extraction from whole human blood based on expanding cell-free layer devices. *Biomed. Microdevices* **2010**, *12*, 485–497.

25. Sajeesh, P.; Sen, A.K. Particle separation and sorting in microfluidic devices: A review. *Microfluid. Nanofluid.* **2014**, *17*, 1–52.

26. Liou, T.M.; Lin, C.T. Study on microchannel flows with a sudden contraction-expansion at a wide range of Knudsen number using lattice Boltzmann method. *Microfluid. Nanofluid.* **2013**, *16*, 315–327.

27. Nejat, A.; Kowsary, F.; Hasanzadeh-Barforoushi, A.; Ebrahimi, S. Unsteady pulsating characteristics of the fluid flow through a sudden expansion microvalve. *Microfluid. Nanofluid.* **2014**, *17*, 623–637.

28. Balan, C.M.; Broboana, D.; Balan, C. Investigations of vortex formation in microbifurcations. *Microfluid. Nanofluid.* **2012**, *13*, 819–833.

29. Liu, S.J.; Wei, H.H.; Hwang, S.H.; Chang, H.C. Dynamic particle trapping, release, and sorting by microvortices on a substrate. *Phys. Rev. E* **2010**, *82*, 026308.

30. Yazdi, S.; Ardekani, A.M. Bacterial aggregation and biofilm formation in a vortical flow. *Biomicrofluidics* **2012**, *6*, 044114.

31. Shang, X.P.; Cui, X.G.; Huang, X.Y.; Yang, C. Vortex generation in a microfluidic chamber with actuations. *Exp. Fluids* **2014**, *55*, 1758.

32. Qin, L.; Vermesh, O.; Shi, Q.; Heath, J.R. Self-powered microfluidic chips for multiplexed protein assays from whole blood. *Lab Chip* **2009**, *9*, 2016–2020.

33. Arifin, D.R.; Yeo, L.Y.; Friend, J.R. Microfluidic blood plasma separation via bulk electrohydrodynamic flows. *Biomicrofluidics* **2007**, *1*, 14103.

34. Yeo, L.Y.; Hou, D.; Maheshswari, S.; Chang, H.C. Electrohydrodynamic surface microvortices for mixing and particle trapping. *Appl. Phys. Lett.* **2006**, *88*, 233512.

35. Fishler, R.; Mulligan, M.K.; Sznitman, J. Mapping low-Reynolds-number microcavity flows using microfluidic screening devices. *Microfluid. Nanofluid.* **2013**, *15*, 491–500.

36. Tsai, C.H.; Yeh, C.P.; Lin, C.H.; Yang, R.J.; Fu, L.M. Formation of recirculation zones in a sudden expansion microchannel with a rectangular block structure over a wide Reynolds number range. *Microfluid. Nanofluid.* **2011**, *12*, 213–220.

37. Tsai, C.H.; Lin, C.H.; Fu, L.M.; Chen, H.C. High-performance microfluidic rectifier based on sudden expansion channel with embedded block structure. *Biomicrofluidics* **2012**, *6*, 24108–241089.

38. Biswas, G.; Breuer, M.; Durst, F. Backward-facing step flows for various expansion ratios at low and moderate Reynolds numbers. *J. Fluids Eng.* **2004**, *126*, 362.

39. Haller, A.; Buchegger, W.; Vellekoop, M. Towards an optimized blood plasma separation chip: Finite element analysis of a novel corner structure in a backward-facing step. *Procedia Eng.* **2011**, *25*, 439–442.

40. Iliescu, C.; Taylor, H.; Avram, M.; Miao, J.; Franssila, S. A practical guide for the fabrication of microfluidic devices using glass and silicon. *Biomicrofluidics* **2012**, *6*, 16505–1650516.

41. Yu, Z.T.F.; Aw Yong, K.M.; Fu, J. Microfluidic blood cell sorting: Now and beyond. *Small* **2014**, *10*, 1687–703.

42. Faivre, M.; Abkarian, M.; Bickraj, K.; Stone, H.A. Geometrical focusing of cells in a microfluidic device: An approach to separate blood plasma. *Biorheology* **2006**, *43*, 147–159.

43. Hou, H.W.; Bhagat, A.A.S.; Lee, W.C.; Huang, S.; Han, J.; Lim, C.T. Microfluidic devices for blood fractionation. *Micromachines* **2011**, *2*, 319–343.

44. Kersaudy-Kerhoas, M.; Sollier, E. Micro-scale blood plasma separation: From acoustophoresis to egg-beaters. *Lab Chip* **2013**, *13*, 3323–3346.

45. Sollier, E.; Rostaing, H.; Pouteau, P.; Fouillet, Y.; Achard, J.L. Passive microfluidic devices for plasma extraction from whole human blood. *Sensors Actuators B Chem.* **2009**, *141*, 617–624.

46. Mach, A.J.; di Carlo, D. Continuous scalable blood filtration device using inertial microfluidics. *Biotechnol. Bioeng.* **2010**, *107*, 302–311.

47. Kersaudy-Kerhoas, M.; Kavanagh, D.M.; Dhariwal, R.S.; Campbell, C.J.; Desmulliez, M.P.Y. Validation of a blood plasma separation system by biomarker detection. *Lab Chip* **2010**, *10*, 1587–1595.

6

Infrared Optical Switch Using a Movable Liquid Droplet

Miao Xu, Xiahui Wang, Boya Jin and Hongwen Ren *

BK Plus Haptic Polymer Composite Research Team, Department of Polymer-Nano Science and Technology, Chonbuk National University, Jeonju, Chonbuk 561-756, Korea;
E-Mails: xumiao0711@sina.com (M.X.); wangxiahui1986@126.com (X.W.); wxffjby@gmail.com (B.J.)

* Author to whom correspondence should be addressed; E-Mail: hongwen@jbnu.ac.kr

Academic Editor: Joost Lötters

Abstract: We report an infrared (IR) optical switch using a wedge-like cell. A glycerol droplet is placed in the cell and its surrounding is filled with silicone oil. The droplet has minimal surface area to volume (SA/V) ratio in the relaxing state. By applying a voltage, the generated dielectric force pulls the droplet to move toward the region with thinner cell gap. As a result, the droplet is deformed by the substrates, causing the SA/V of the droplet to increase. When the voltage is removed, the droplet can return to its original place in order to minimize the surface energy. Owing to the absorption of glycerol at 1.55 μm, the shifted droplet can be used to attenuate an IR beam with the advantage of polarization independent. Fluidic devices based on this operation mechanism have potential applications in optical fiber switches, IR shutter, and variable optical attenuations.

Keywords: liquid droplet; dielectric force; optical switch; attenuation ratio

1. Introduction

Microfluidics is a technology that enables precise, automated manipulation of tiny volume of a fluid. With the rapid growth of microfluidic technology, microfluidic devices have attracted tremendous interest for the development of lab-on-a-chip systems [1–3]. A variety of methods have been proposed for the fabrication of microfluidic devices. These devices have widespread applications in micro-pumping [4,5], inkjet printing [6], sensing [7], biology [8,9], beam steering [10,11], and adaptive lenses [12–16]. Due to optical isotropy, microfluidic has also been used for optical switches. Examples include displays [17,18] and adaptive iris [19–23]. Most of these approaches are used to switch visible light, but few are suitable for switching infrared (IR) light.

An IR light switch is a device used to reduce the optical power at a certain level, either in free space or in an optical fiber. It has been an essential component in optical fiber switching, photonic signal processing, IR shutter, attenuators, and sensing. To electrically switch an IR light, modulators based on light scattering [24,25], phase modulation [26,27], beam steering [10], and light absorption [28] have been proposed. Each approach has its own strengths and limitations. A liquid crystal (LC) device can attenuate IR light by either scattering or phase modulation without moving parts. For the scattering type, the LC usually forms droplets in a polymer matrix. When an unpolarized light passes through the LC device, the output light is sensitive to polarization because the droplets are optically anisotropic. For the phase type using a blue phase LC, it usually provides a rather limited phase shift [27,28]. Increase the cell gap can enhance the phase shift, but the driving voltage will increase dramatically. A liquid device based on either electrowetting or dielectrophoretic effect can be used for modulating an IR beam. In an electrowetting device, salty water is commonly used as the key conductive liquid [10,12,13]. By deforming the shape of a water droplet using electrostatic force, light is controlled when it passes through the droplet. Because water is transparent to IR light, optical modulation is obtained only by surface deflection. As a comparison, a dielectrophoretic device may not use deionized water as the liquid. Instead, one can choose a suitable dielectric liquid which not only has good physical properties, but also can strongly absorb an IR light. By shifting the position of the dielectric liquid, an optical switch can be obtained. Such an optical switch is much attractive because the liquid functions as an adaptive neutral density filter. In a previous report [29], a glycerol droplet is used for an IR optical switch because glycerol strongly absorbs IR light at 1.55 μm. Due to the special cell structure, the glycerol droplet could do a rather limited displacement. As a result, it could not yield a sufficient space for a large area optical switch.

Here we report a liquid-based optical switch using a wedge-like cell. A glycerol droplet is placed in the cell and its surrounding is filled with silicone oil. The electrode on both substrate surfaces is continuous. When the droplet is actuated electrically, it can present a reciprocal motion. The shifted droplet can switch an IR beam at 1.55-μm wavelength. For the cell with 6.5-mm-diameter droplet, the attenuation ratio can reach ~110:1. The optical switch is polarization independent. Similar to previous fabrication [30], our device can be prepared easily without patterning the electrode and the cell gap is not required to be controlled precisely. Moreover, our device owns the advantages of simple structure and direct voltage actuation.

2. Cell Structure and Mechanism

The side-view structure and the operation mechanism of the liquid cell are schematically depicted in Figure 1a. Two Indium-Tin-Oxide (ITO) glass substrates are placed together with a tilt angle. The surface of each substrate has a hydrophobic layer. The hydrophobic layer has two functions: provides low surface tension and prevents charges injection from the electrodes. Two liquids are sandwiched by the substrates. The two liquids are immiscible and have different dielectric constants. One liquid forms a droplet and the other is used to fill its surrounding. In the relaxing state, the droplet slightly touches both substrates with minimal surface area to volume (SA/V) ratio. By applying a voltage, the generated dielectric force can pull the droplet to move toward the region with thinner gap, as shown in Figure 1b.

Because the volume of the droplet is fixed, the droplet is further deformed by the substrates, causing its *SA/V* to increase. If the voltage is sufficiently high and the length of the cell in horizontal position is much larger than the diameter of the droplet, then the droplet can shift with a large displacement. Once the voltage is removed, the droplet will return to its original place in order to minimize the surface energy.

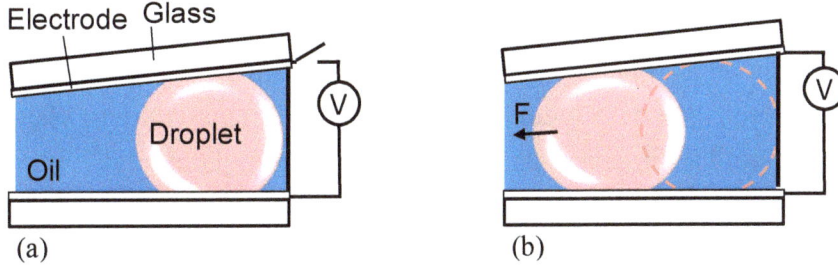

Figure 1. Cross-sectional structure of the cell in (**a**) relaxing state and (**b**) actuating state.

To depict the moving mechanism of the droplet in the actuating state, a cross-sectional structure with defined parameters is given in Figure 2. The droplet has four contact angles on the substrate surfaces. The contact angle (θ) is defined geometrically as the angle formed by the droplet at the three-phase boundary where droplet, medium liquid, and substrate surface intersect. If the effect of gravity on deforming the droplet is neglected, then the contact angle satisfies $\theta_A = \theta_B < \theta_C = \theta_D$. This is because the bending of the left part of the droplet is severer than that of the right part. In the actuating state, the molecules at the border of the droplet experience the strongest dielectric force. The dielectric force is expressed by [31]:

$$F = -\frac{1}{2}\varepsilon_0(\varepsilon_1 - \varepsilon_2)\nabla E^2 \qquad (1)$$

where ε_0 represents the permittivity of free space, ε_1 and ε_2 represent the dielectric constants of the droplet and the surrounding liquid, respectively, and ∇E denotes the gradient of the electric field on the droplet. In this cell, the dielectric constant of the droplet is larger than that of the surrounding liquid. The electric field near′the edge of the droplet is perpendicular to the droplet surface. Therefore, **F** is perpendicular to the droplet surface too. According to Equation (1), the droplet at points A, B, C, and D experience the forces F_A, F_B, F_C, and F_D, respectively. Each force can be resolved into a component along the substrate surface with the relationship of $F_{A1} = F_{B1} > F_{C1} = F_{D1}$, therefore, the resultant force can pull the droplet to move toward the region with thinner cell gap. When the interfacial tensions and the dielectric force are balanced, the droplet will stop to move.

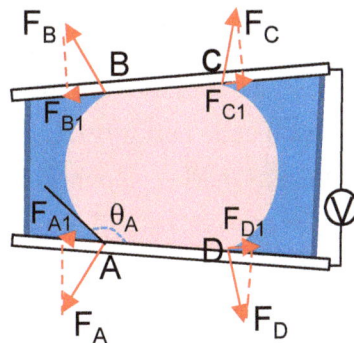

Figure 2. Mechanism of the droplet actuated by the generated dielectric force.

3. Cell Fabrication

To prepare a liquid cell as shown in Figure 1a, two ITO glass substrates are chosen and their surfaces are coated with a thin Teflon layer (400S1-100-1, from DuPont, Wilmington, DE, USA). Teflon is a desired hydrophobic material with a low surface tension ($\gamma_T \sim 18$ mN/m) at room temperature. We then use glass stripes to build a fence on the boundary of one substrate in order to form a tiny container. The fence has a slope with $\sim 5°$ tilt angle. The largest thickness of the glass strip is ~ 2.3 mm and the smallest thickness is ~ 0.5 mm. Glycerol ($\rho_g \sim 1.25$ g/cm^{-3}, purity $\geq 99.5\%$, Sigma-Aldrich, Malaysia) is chosen as the droplet because glycerol has three unique properties: absorbs light around 1.55-μm wavelength, has a large dielectric constant ($\varepsilon_g \sim 47$), and has a large surface tension ($\gamma_g \sim 63$ mN/cm). For easy observation, the glycerol is doped with ~ 0.1 wt % Rose Bengal (Sigma-Aldrich). A small amount of the glycerol is dripped on the bottom substrate. Silicone oil ($\varepsilon_2 \sim 2.9$, $\gamma_S \sim 21$ mN/m, $\rho \sim 0.97$ g/cm^{-3}) is chosen to fill its surrounding as the medium. The container is then covered with the other glass substrate to form a cell. The periphery of the cell is tightly sealed using epoxy glue. Because the diameter of droplet is larger than the cell gap, the droplet is squeezed by the two substrates and shifts to the region with thicker gap. In the relaxing state, the droplet has the smallest SA/V ratio. The diameter of the droplet is measured to be ~ 6.5 mm.

4. Results and Discussion

To evaluate the absorption of the two liquids to IR light, the transmission spectra of the two liquids (2.3-mm thick) is measured. The results are given in Figure 3. In the range of 1.45–1.6 μm wavelength, glycerol is opaque, while silicone oil is highly transparent.

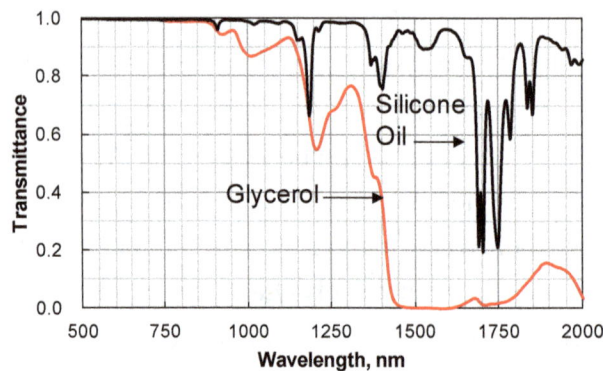

Figure 3. Transmission spectra of glycerol and silicone oil with 2.3-mm thick.

Although glycerol molecules can highly absorb IR light around 1.55-μm wavelength, the transmittance of a glycerol layer is dependent on its thickness. To measure the transmittance of glycerol, several cells with different gaps are prepared. The gap of each cell is uniform. Each cell is filled with glycerol. An IR laser beam (LAS DFB-1550-6, $\lambda = 1.55$ μm, Laser Max, Rochester, NY, USA) is normally incident upon the cell. The transmitted light intensity is received by a photodiode. The transmitted intensity with different cell gaps is shown in Figure 4. Increase the cell gap can decrease the transmitted intensity. When the cell gap reaches ~ 2.3 mm, the laser beam is highly blocked with the lowest transmittance. According to this result, the largest thickness of our wedge cell is controlled to be ~ 2.3 mm.

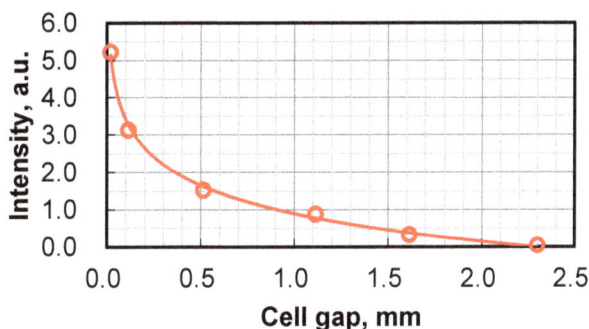

Figure 4. Thickness of glycerol layer *versus* the transmitted intensity of a laser beam.

To measure the travel distance of the droplet when it is actuated by a voltage, a ruler is placed under the cell. It is convenient to use a digital microscope to record the droplet. The droplet presents red color due to the doped Rose Bengal. When $V = 0$, a round droplet is observed, as shown in Figure 5a. The diameter of the spot is ~6.5 mm. When the voltage is gradually increased to 40 V_{rms}, the droplet begins to shift, implying that the generated dielectric force breaks the balance of the interfacial tensions. When $V = 40$ V_{rms}, the droplet can travel ~1.5 mm, as shown in Figure 5b. When $V = 50$ V_{rms}, the droplet can move in the same direction and has ~7 mm displacement. As the droplet is pulled to move toward the region with thinner gap (left), the droplet is squeezed by the two substrates, causing its SA/V to increase, as shown in Figure 5c. When the voltage is removed, the droplet returns to its initial position in order to minimize its surface energy, as shown in Figure 5d. It takes ~30 s for the droplet to do one reciprocal motion. The slow response time is due to the long travel distance of the droplet. Moreover, the viscosity of the glycerol also affects the response time.

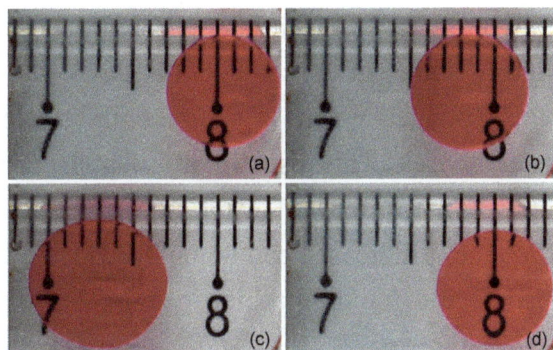

Figure 5. The motion of the droplet impacted with different voltages. (**a**) $V = 0$, (**b**) $V = 40$ V_{rms}, (**c**) $V = 50$ V_{rms}, and (**d**) after removing the voltage for 17 s.

Actuating with different voltages, the droplet can take different times when it travels the same distance and returns to its original place. The duration time is the time when a constant voltage is used to pull the droplet to travel a certain distance. The returning time is the time the droplet spends when it returns to its original place. Figure 6 shows the duration/returning time *versus* different voltages when the droplet travels 7-mm distance. Increase the voltage can decrease the duration time, but the returning time has a tendency to increase. Increase the voltage can enhance the gradient of electric field (∇E) and electric field (E) between the two substrates. From Equation (1) and Figure 2, the generated dielectric force can accelerate the moving speed of the droplet, thus decreasing the duration time. If the droplet is pulled to move ~7 mm with a higher speed, then the droplet will keep this inertia once the voltage is removed. The inertia will cause the droplet to be deformed largely. The droplet needs some time to recover to its shape and then returns back. Therefore, an increased voltage impacted on the droplet can cause the droplet to spend more time to return back.

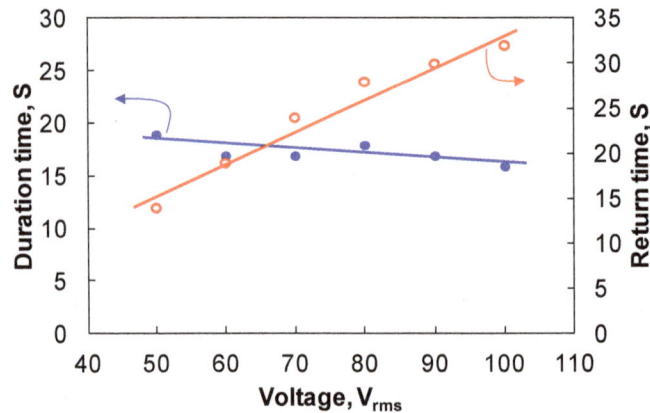

Figure 6. Duration time and returning time when the droplet is pulled to move 7 mm with different voltages.

Because the droplet is big enough to block a thin IR beam, sometimes it is not necessary to largely shift the droplet when it is used for an optical switch. To measure the optical switch of the droplet as shown in Figure 1, the diameter of the collimated laser beam is controlled to be ~1.5 mm. In the relaxing state, the probing beam is blocked by the right part of the droplet, as shown in Figure 7a. When a voltage is applied to the cell, the droplet is pulled to shift and yields some space so that the laser beam can completely pass through the silicone oil, as shown in Figure 7b. The transmitted light is received using the photodiode. To accelerate the moving speed of the droplet, a compromised method is to increase the driving voltage and decrease the duration time of the voltage. Figure 7c shows the dependence of the transmitted intensity on the applied voltage. When $V = 0$, the light intensity is minimal. When $V = 30$ V_{rms}, the transmittance starts to increase. This result implies that the droplet moves to the left and partial beam passes through the cell. When $V = 87$ V_{rms}, the light intensity is the highest. The light intensity is saturated although the voltage is increased continuously. This result implies that the droplet has completely yielded its occupied region for the probing beam. The maxima attenuation ratio is measured to be ~110:1.

As an IR shutter, the response time of the droplet in the cell should be evaluated. Similar to the method measured in Figure 7, the switching time is evaluated by monitoring the time dependent transmitted light intensity. A pulse voltage instead of a continuously changed voltage is used to drive the droplet. The transmitted light intensity is detected by the photodiode and recorded using an oscilloscope. To measure the response time, a 100-V_{rms} pulse voltage (500 Hz) is applied to the cell. The duration time is 13 s. The result is shown in Figure 8. It takes ~6 s for the droplet to yield the place for the probing beam. When the voltage pulse is removed, the transmitted intensity firstly keeps constant for a while (~1.5 s), then begins to decrease. This implies that the travel distance of the droplet is larger than the diameter of the beam. Only when the right border of the droplet meets the beam, the intensity starts to decrease. It takes ~7 s for the droplet to return to its original position. Three cycles show that the droplet can repeat very well for the reciprocating movement. To precisely control the travel distance of the droplet for an optical switch, a feasible method is to adjust the duration time of the voltage.

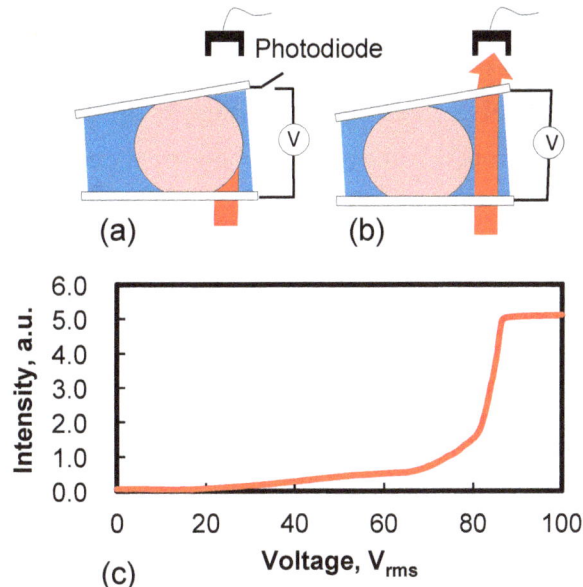

Figure 7. Method to measure the electro-optical property of the liquid cell. (**a**) Blocking the probing beam, (**b**) uncovering the beam, and (**c**) dependent of light transmission on the driving voltage. The time of the voltage applying on the droplet is 17 s.

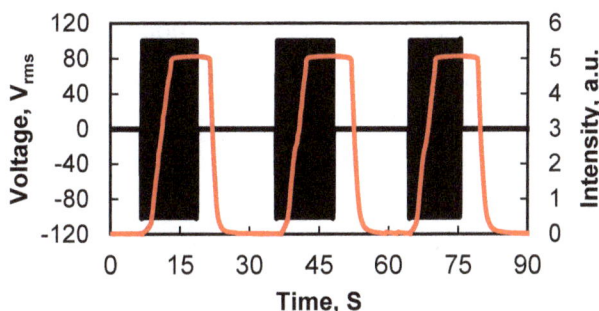

Figure 8. Light intensity change impacting with 100 V_{rms} voltage pulse.

In contrast to previous liquid device [29], our droplet can make a large displacement in its cell. This is because the dielectric layer coated on the two substrates is not specially treated, as depicted in Figure 2. The tradeoff is the slow response time. For a short displacement, the response time can be reduced. Due to the wedge-like structure, a droplet with different volumes can be used to fill the cell, so the droplet can be easily scaled up for different applications, such as IR shutters, optical fiber switches, and optical attenuators. Although the densities of the two liquids do not match well, the gravity effect will not affect the optical switching of the droplet when the cell is placed in vertical position. This is because both substrates always sustain the droplet whether or not it is actuated. For example, when the cell is placed in vertical position, the boundary of the cell may participate to sustain the droplet as well. Therefore, the cell has good mechanical stability without the issue of shocking or vibrating.

The liquid cell works very well to attenuate an IR beam at 1.55 μm. Different from a device which controls a beam by scattering or steering, this device can obtain a high attenuation ratio without degrading the performance of the incident light. Although the liquid cell absorbs IR light, the liquid still can present good thermal stability [32]. Moreover, the surrounding medium can help dissipate the produced heat. Therefore, it is not an issue when the cell is used to switch an IR beam. By choosing proper liquids in the desired spectral region, the device can be used to switch mid-wavelength IR and long-wavelength IR. To improve the returning time of the droplet, some criteria of the liquids, e.g., viscosity, interfacial tensions, and the size of the droplet should be considered. For an optical fiber switch, the device can work in reflective mode if the surface of the top substrate is coated with a reflector. In the reflective mode, the droplet can obtain the highest attenuation ratio because the droplet absorbs the beam for two times. Therefore, this device has potential applications in optical communication, variable optical attenuators, and other lab-on-a-chip systems.

5. Conclusions

We have reported an IR light switch using a movable glycerol droplet. The droplet is sealed in a wedge cell and its surrounding is filled with silicone oil. The droplet could do a reciprocal motion driven by the generated dielectric force. Due to the absorption of light at 1.55 μm, the droplet can be used for an optical switch. For the glycerol droplet with 6.5-mm diameter, the attenuation ratio can reach ~110:1 in the transmissive mode. The response time of the motion is dependent on the amplitude of the applied voltage. A higher voltage can accelerate the moving speed of the droplet, but the returning speed is slower instead. Due to the large viscosity of the glycerol, the response time of the glycerol is slow when it has a large displacement. For a short travel distance, this issue can be solved by choosing a suitable liquid pair. Our demonstrated device owns the advantages of large displacement, scalable droplet size, and good mechanical stability. If the device is used in reflective mode, then the attenuation ratio will be enhanced largely. Fluidic devices based on our cell structure have potential applications in IR light shutter, fiber-optic switches, and lab-on-a-chip systems.

Acknowledgments

This work is financially supported by the National Research Foundation of Korea under Grant 2014001345 and partially supported by the Basic Science Research Program of NRF under Grant 2014064156.

Author Contributions

All authors helped conceive the idea and prepared the manuscript.

Conflicts of Interest

The authors declare no conflict of interest.

References

1. Monat, C.; Domachuk, P.; Eggleton, B.J. Integrated optofluidics: A new river of light. *Nat. Photon.* **2007**, *1*, 106–114.
2. Levy, U.; Shamai, R. Tunable optofluidic devices. *Microfluid. Nanofluid.* **2008**, *4*, 97–105.
3. Xu, S.; Ren, H.W.; Wu, S.T. Dielectrophoretically tunable optofluidic devices. *J. Phys. D Appl. Phys.* **2013**, *46*, doi:10.1088/0022-3727/46/48/483001.
4. Malouin, B.A., Jr.; Vogel, M.J.; Olles, J.D.; Cheng, L.L.; Hirsa, A.H. Electromagnetic liquid pistons for capillarity-based pumping. *Lab Chip* **2011**, *11*, 393–397.

5. Ren, H.W.; Xu, S.; Wu, S.T. Liquid crystal pump. *Lab Chip* **2013**, *13*, 100–105.

6. Boland, T.; Xu, T.; Damon, B.; Cui, X.F. Application of inkjet printing to tissue engineering. *Biotechnol. J.* **2006**, *1*, 910–917.

7. Wu, J.; Gu, M. Microfluidic sensing: State of the art fabrication and detection techniques. *J. Biomed. Opt.* **2011**, *16*, doi:10.1117/1.3607430.

8. Breslauer, D.N.; Lee, P.J.; Lee, L.P. Microfluidics-based systems biology. *Mol. BioSyst.* **2006**, *2*, 97–112.

9. Chung, B.G.; Lee, K.H.; Khademhosseini, A.; Lee, S.H. Microfluidic fabrication of microengineered hydrogels and their application in tissue engineering. *Lab Chip* **2012**, *12*, 45–59.

10. Reza, S.A.; Riza, N.A. A liquid lens-based broadband variable fiber optical attenuator. *Opt. Commun.* **2009**, *282*, 1298–1303.

11. Lin, Y.J.; Chen, K.M.; Wu, S.T. Broadband and polarization-independent beam steering using dielectrophoresis-tilted prism. *Opt. Express* **2009**, *17*, 8651–8656.

12. Krupenkin, T.; Yang, S.; Mach, P. Tunable liquid microlens. *Appl. Phys. Lett.* **2003**, *82*, 316–318.

13. Kuiper, S.; Hendriks, B.H.W. Variable-focus liquid lens for miniature cameras. *Appl. Phys. Lett.* **2004**, *85*, 1128–1130.

14. Liang, D.; Agarwal, A.K.; Beebe, D.J.; Jiang, H. Adaptive liquid microlenses activated by stimuli-responsive hydrogels. *Nature* **2006**, *442*, 551–554.

15. Cheng, C.C.; Yeh, J.A. Dielectrically actuated liquid lens. *Opt. Express* **2007**, *15*, 7140–7145.

16. Ren, H.W.; Xianyu, H.Q.; Xu, S.; Wu, S.T. Adaptive dielectric liquid lens. *Opt. Express* **2008**, *16*, 14954–14960.

17. Hayes, R.A.; Feenstra, B.J. Video-speed electronic paper based on electrowetting. *Nature* **2003**, *425*, 383–385.

18. Xu, S.; Ren, H.; Liu, Y.F.; Wu, S.T. Color displays based on voltage-stretchable liquid crystal droplet. *J. Display Technol.* **2012**, *8*, 336–340.

19. Murade, C.U.; Oh, J.M.; Van den Ende, D.; Mugele, F. Electrowetting driven optical switch and tunable aperture. *Opt. Express* **2011**, *19*, 15525–15531.

20. Tsai, C.G.; Yeh, J.A. Circular dielectric liquid iris. *Opt. Lett.* **2010**, *35*, 2484–2486.

21. Zuta, Y.; Goykhman, I.; Desiatov, B.; Levy, U. On-chip switching of a silicon nitride micro-ring resonator based on digital microfluidics platform. *Opt. Express* **2010**, *18*, 24762–24771.

22. Ren, H.W.; Xu, S.; Wu, S.T. Voltage-expandable liquid crystal surface. *Lab Chip* **2011**, *11*, 3426–3430.

23. Li, L.; Liu, C.; Wang, Q.H. Optical switch based on tunable aperture. *Opt. Lett.* **2012**, *37*, 3306–3308.

24. Takizawa, K.; Kodama, K.; Kishi, K. Polarization-independent optical fiber modulator by use of polymer-dispersed liquid crystals. *Appl. Opt.* **1998**, *37*, 3181–3190.

25. Hirabayashi, K.; Wada, M.; Amano, C. Compact optical-fiber variable attenuator arrays with polymer network liquid crystals. *Appl. Opt.* **2001**, *40*, 3509–3517.

26. Chanclou, P.; Vinouze, B.; Roy, M.; Cornu, C. Optical fibered variable attenuator using phase shifting polymer dispersed liquid crystal. *Opt. Commun.* **2005**, *248*, 167–172.

27. Zhu, G.; Wei, B.Y.; Shi, L.Y.; Lin, X.W.; Hu, W.; Huang, Z.D.; Lu, Y.Q. A fast response variable optical attenuator based on blue phase liquid crystal. *Opt. Express* **2013**, *21*, 5332–5337.

28. Lin, Y.-H.; Chen, H.-S.; Lin, H.-C.; Tson, Y.-S.; Hsu, H.-K.; Li, W.-Y. Polarizer-free and fast response microlens arrays using polymer-stabilized blue phase liquid crystals. *Appl. Phys. Lett.* **2010**, *96*, doi:10.1063/1.3360860.

29. Ren, H.; Xu, S.; Liu, Y.; Wu, S.T. Liquid-based infrared optical switch. *Appl. Phys. Lett.* **2012**, *101*, doi:10.1063/1.4738995.

30. Smith, N.R.; Abeysinghe, D.C.; Haus, J.H.; Heikenfeld, J. Agile wide-angle beam steering with electrowing microprisms. *Opt. Express* **2006**, *14*, 6557–6563.

31. Penfield, P.; Haus, H.A. *Electrodynamics of Moving Media*; MIT Press: Cambridge, MA, USA, 1967.

32. Zhang, H.; Ren, H.; Xu, S.; Wu, S.T. Temperature effects on dielectric liquid lenses. *Opt. Express* **2014**, *22*, 1930–1939.

7

Multiplex, Quantitative, Reverse Transcription PCR Detection of Influenza Viruses Using Droplet Microfluidic Technology

Ravi Prakash [1], Kanti Pabbaraju [2], Sallene Wong [2], Anita Wong [2], Raymond Tellier [2,3] and Karan V. I. S. Kaler [1,*]

[1] Biosystems Research and Applications Group, Department of Electrical and Computer Engineering, Schulich School of Engineering, University of Calgary, Calgary, AB T2N 1N4, Canada; E-Mail: rprakash@ucalgary.ca

[2] Provincial Laboratory for Public Health of Alberta, Calgary, AB T2N 4W4, Canada; E-Mails: kanti.pabbaraju2@albertahealthservices.ca (K.P.); sallene.wong@albertahealthservices.ca (S.W.); anita.wong@albertahealthservices.ca (A.W.); raymond.tellier@albertahealthservices.ca (R.T.)

[3] Department of Microbiology, Immunology and Infectious Diseases, Cumming School of Medicine, University of Calgary, Calgary, AB T2N 1N4, Canada

* Author to whom correspondence should be addressed; E-Mail: kaler@ucalgary.ca

Academic Editor: Joost Lötters

Abstract: Quantitative, reverse transcription, polymerase chain reaction (qRT-PCR) is facilitated by leveraging droplet microfluidic (DMF) system, which due to its precision dispensing and sample handling capabilities at microliter and lower volumes has emerged as a popular method for miniaturization of the PCR platform. This work substantially improves and extends the functional capabilities of our previously demonstrated single qRT-PCR micro-chip, which utilized a combination of electrostatic and electrowetting droplet actuation. In the reported work we illustrate a spatially multiplexed micro-device that is capable of conducting up to eight parallel, real-time PCR reactions per usage, with adjustable control on the PCR thermal cycling parameters (both process time and temperature set-points). This micro-device has been utilized to detect and quantify the presence of two clinically relevant respiratory viruses, Influenza A and Influenza B, in human samples (nasopharyngeal swabs, throat swabs). The device performed accurate detection and quantification of the two respiratory viruses, over several orders of RNA copy counts, in unknown (blind) panels of extracted patient samples with acceptably high PCR efficiency (>94%). The multi-stage qRT-PCR assays on eight panel patient samples were accomplished within 35–40 min, with a detection limit for the target Influenza virus RNAs estimated to be less than 10 RNA copies per reaction.

Keywords: reverse transcription PCR (RT-PCR); Influenza viruses; droplet microfluidics (DMF); nano-texture; dielectrophoresis (DEP); electrostatics; electrowetting (EW); multiplex qRT-PCR assays

1. Introduction

The detection of clinically relevant viral pathogens is an essential task performed by medical microbiology laboratories, to help establish diagnosis, guide the subsequent treatment and contribute to public health surveillance, including monitoring of emerging agents. The introduction of molecular detection methods and especially nucleic acid amplification techniques, such as polymerase chain reaction (PCR) [1,2], has revolutionized the capabilities of diagnostic virology laboratories in large part due to the increased sensitivity of detection and improved turnaround time. Since different viruses can have similar clinical presentations, patients typically have to be tested for the presence of several different viruses at the same time; this is leading increasingly to the setting up of testing panels for several viruses linked to a specific syndrome. Some typical examples include gastroenteritis virus panels, viral encephalitis panels and of course respiratory virus panels. In practice, "panels" can be performed in two different ways; several different assays on a single sample with multiple targets/markers by using spectral multiplexing or, by utilizing spatial multiplexing based parallel PCR assays where separate individual sample and target mixtures are prepared and amplified in parallel [3].

PCR is used in the screening and detection of numerous infectious viral or bacterial species, by amplifying the target nucleic acids extracted from patient sample, over several orders of magnitude [1,2]. Quantitative polymerase chain reaction (qPCR) furthermore facilitates real-time detection of target nucleic acid during the amplification process and allows for quantitation of the initial amount of template [4,5]. PCR at microscale leads to a reduction in the reaction time, bio-sample/reagent volume [6,7]. Following the early development of the conventional close channel microfluidics based PCR micro-devices [6–9], handling the PCR sample volume in form of rapidly dispensed, discrete microliter or smaller droplets has become a preferred method of choice for PCR micro-devices [10,11]. Most microfluidic PCR systems have focused primarily on either reducing the PCR reaction volume (down to few hundred nanoliter) [12–14] or reducing the PCR reaction time (~10 min) [12–16]. The requirement of high surfactant concentration in the continuous oil phase to stabilize the PCR droplets, lack of individual addressing of the multiple droplets and the excessive need for off-chip overhead (pumps, plumbing, valves) are just a few of the challenging issues driving the development of miniaturized PCR set-ups towards droplet microfluidic (DMF). Among the available DMF methods, electro-actuation of droplets on patterned substrates (Glass, Silicon, Polymer, *etc.*) is the most effective means of precision dispensing and subsequently handling multitude of bio-sample and reagents using a miniaturized device. Electrowetting (EW) or, EW-on-dielectric (EWOD) [17,18] has been successfully utilized to demonstrate PCR reactions at microscale [19–22], however, the necessity of active electrode switching to facilitate droplet motion, in the digital microfluidic technique, results in a substantial electrical overhead, especially for the multiplexed, chip based bio-assay schemes. Recently, our work has demonstrated that a continuous droplet transport scheme, which enables droplet transport and thermal cycling without the requirement of active electrode switching [11], can be an effective solution for a PCR micro-device with reduced electrical overhead requirement for a multiplexed diagnosis system. Here active droplet transport is facilitated by electrostatic or, droplet-dielectrophoresis (D-DEP) based electro-actuation technique, which utilizes herring-bone shaped electrode arrays to facilitate droplet transport and thermal cycling [11,23,24].

In cases pertaining to the detection and quantification of RNA viruses, for example Influenza viruses, Hepatitis C virus, Measles virus, SARS-CoV and Ebola virus, detection by PCR requires transcribing the viral RNA extracted from virions through a reverse transcription reaction, to yield complementary DNA (cDNA) molecules. Apart from virology, other major applications of RT-PCR include analysis of gene expression from target cells and detection of certain genetic diseases. In quantitative analysis of RT-PCR (qRT-PCR), a reaction mixture containing both a reverse transcriptase enzyme and a thermo-stable DNA polymerase (TAQ) is used so that the two enzymatic reactions (reverse transcription and PCR amplification) can be performed serially through temperature control, as an integrated two-step process. On a miniaturized scale, RT-PCR reactions have been achieved by utilizing microfluidic methods for manipulating nucleic acid samples and PCR reagents including the use of continuous flow techniques [6–9] and discrete droplet based microfluidics [10,11]. The majority of microfluidic implementations for RT-PCR assays have been targeted to gene expression analysis, where a large amount of genomic molecules has helped towards lowering the reaction volumes to sub-microliters [12–14]. However, in clinical diagnostic applications, for the detection of trace quantities of viral RNA in a matrix sample that often contains an abundance of genomic DNA from the human host, the emphasis is placed on the reliable, rapid detection [15].

The evolution of PCR technologies over the last two decades suggests the need for further improvement towards the performance and reliability of PCR systems. The droplet digital PCR (ddPCR™, Bio-Rad, Hercules, CA, USA) system commercially available from Bio-Rad incorporates the close-channel microfluidic based droplet generation method to create a large library (~15,000) of sub-microliter droplets, which are dispensed from a large PCR sample/reagent mixture (~20 μL) [16]. This approach allows for one step detection and quantification of extracted nucleic acid and it is an excellent example of PCR technology that illustrates the feasible integration of microfluidics into such commercial systems.

Reports of large PCR sample arrays of sub-microliter reactions have also been reported for high-throughput qRT-PCR, with as few as five copies of template RNA in each reaction [14]. However, detection of a panel of infectious viruses (such as the respiratory virus panel) in human samples utilizing a qRT-PCR microchip remains to be realized. Examples of recent single RT-PCR applications using microfluidics suggest a detection time (reverse transcription and thermal cycling) of up to an hour, using as low as 2–5 μL PCR volume [14,15]. We have previously designed a microchip utilizing electrostatic/droplet-DEP (D-DEP) electro-actuation method and integrated thermostatic zones (micro-heaters and resistive temperature sensors) to achieve single qRT-PCR amplification of *in vitro* synthesized Influenza viral RNA [11], with a detection threshold of less than 10 copies of template RNA in the PCR reaction volume. We have also investigated the scalability of PCR sample volume in our device application, over the range of 1–10 μL [11], which is industrially accepted for viral detection in clinical samples. In this work, we have modified the previously designed continuous,

D-DEP electrode architecture for the PCR thermal cycling to produce a spatially multiplexed PCR micro-device, suitable for carrying out several different qRT-PCR reactions in parallel (up to eight assays per chip) and with a built-in flexibility to accommodate different cycling parameters for each reaction. The performance of this micro-device is illustrated by the parallel execution of assays for the detection of Influenza A virus and Influenza B virus in different panels of clinical samples. The reported multiplexed qRT-PCR assays are a first demonstration of a D-DEP based DMF device for analysing multiple clinically relevant viral pathogens in panels of extracted patient samples.

2. Experimental Section

2.1. Device Fabrication

The micro-device was designed using the MEMSPro L-Edit (v. 8.0) CAD software. The DMF chip was fabricated at a micro/nano fabrication facility (Nanofab, Edmonton, Canada). The detailed fabrication method has been previously reported [11,25]. The fabrication procedure utilized to produce the micro-device are identical to the one used in the development of single qRT-PCR microchips, shown in Figure 1a [11]. A pair of 6.6 cm × 3.0 cm micro-devices was fabricated from a 10 cm square glass (Borofloat) wafer (Figure 1b,c). The micro-device consists of: (1) an array of photo lithographically patterned chromium (Cr thickness: 200 nm) micro-heaters and resistance temperature detectors (RTDs) to create the two thermostatic zones (Heater blocks 1 and 2) required during the thermal cycling, (2) a photo lithographically patterned gold/chrome overlay (100 nm Au/ 200 nm Cr) for electrical connections to the micro-heaters/RTD sensors, (3) another photo lithographically patterned Aluminum (200 nm) layer for D-DEP electrodes and (4) Au/Cr metallization for the EW track, utilized for loading the PCR template and reagent mix droplets to the thermal cycler electrodes. These 3 different metal layers were electrically isolated and passivated using dielectric stacks of silicon nitride (Si_3N_4 thickness: 500 nm), to prevent sample electrolysis during electro-actuations. The very top dielectric layer was furthermore utilized to produce a nano-textured super hydrophobic (SH) top surface, utilizing a soft lithography technique [25]. The SH surface provided a high droplet contact angle (CA ~156°) during the device application and significantly minimizes bio-sample adsorption [11,25].

Figure 1a demonstrates the first generation single qRT-PCR microchip, which facilitates spiral droplet transport between the two heater blocks [11]. This electrode structure required up to 10 s to convey the droplet from one thermal zone to another. This delay is principally due to the two relay-controlled track switching required to facilitate the spiral droplet transport. While attempting to improve the electrode architecture towards a more compact single cell design which can result in a larger assay matrix from a 10 cm substrate, it was observed that a single bi-direction track (see Figure 1b significantly reduces the PCR cell area by up to 25% and facilitates droplet transport from one zone to another in ~5 s with one track switching (on the end). This coupled with the 25–30 s annealing period (in the lower temperature zone) ensured the reduction of droplet track size and hence the thermal cycling time. Similar to our previous work, a standard fluorescent thermometry dye (Rhodamine B dye) was used to verify the temperature of the droplet during the annealing and denaturation phase of the PCR thermal cycle [26,27]. Figure 1b shows the improved bi-direction electrode structure as part of the multiplexed (eight-plex) qRT-PCR unit (Figure 1c), which was fabricated and utilized in all the experimental work, reported in this paper.

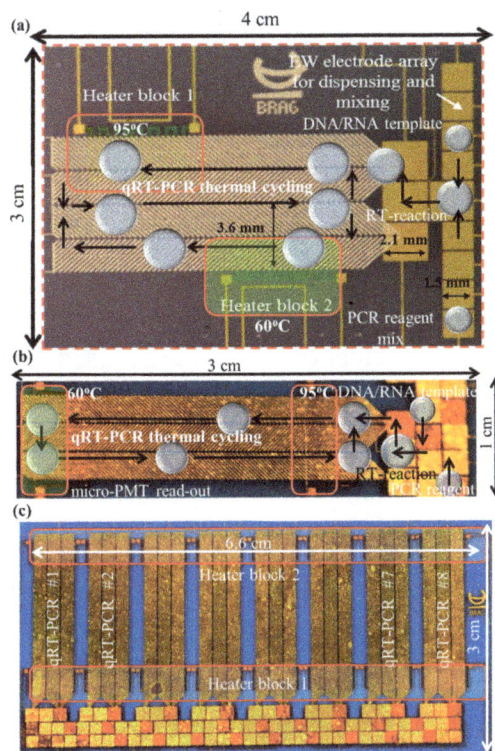

Figure 1. (**a**) Photomicrographs of (**a**) the spiral droplet-dielectrophoresis (D-DEP) electrode architecture used in the earlier single quantitative, reverse transcription, polymerase chain reaction (qRT-PCR micro-device) [11]; (**b**) the continuous, bi-directional droplet actuation scheme and; (**c**) the eight-plex micro-device.

2.2. Sample Preparation

The various sample preparation protocols used in this work are detailed below.

a. Extraction of total nucleic acid from clinical specimens. The extracted nucleic acids, including RNA, were from left-over samples from patients, initially submitted to ProvLab for Influenza virus detection; nucleic acid extracts from samples were labeled at ProvLab as positive for Influenza A or Influenza B or negative, but were otherwise anonymized. Initially, respiratory samples including nasopharyngeal swabs (NP) and throat swabs (TS) were pre-treated with 25 μL of 0.01 mAU/μL of protease (Qiagen, Mississauga, Ontario, Canada) in a thermomixer (Eppendorf, Westbury, NY, USA) at 56 °C and 1000 rpm for 10 min and the supernatant was collected for the extraction process. The total nucleic acid was extracted from the treated samples using the easyMAG® automated extractor (bioMérieux, Montreal, Canada) according to the manufacturer's instructions [11]. The extracted nucleic acid was eluted into a final volume of 110 μL of elution buffer (Borate buffer; pH 8.5) from a sample input volume of 200 μL.

b. qRT-PCR assay. All samples used for validation studies underwent extraction and were tested for Influenza A and Influenza B using real-time RT-PCR assays. The primer and probe sequences from previously reported real-time RT-PCR assays (developed at the Center for Disease Control (CDC), USA) were used for the detection of Influenza A and Influenza B viral RNA. The Influenza A assay targets the matrix gene and the Influenza B assay targets the non-structural gene resulting in the amplification of a 105 base pair product for influenza A and 103 base pair product for Influenza B.

Amplification was performed by one-step RT-PCR using the TaqMan® Fast Virus One-Step RT-PCR Master Mix (Life Technologies Inc., Burlington, Canada), 0.8 μM each of sense and antisense primers and 0.2 μM of the labeled probe. Five microliters of *in vitro* RNA was combined with 5 μL of the master mix. The reaction parameters included a reverse transcription (RT) step performed at 50 °C for 5 min, followed by enzyme activation at 95 °C for 20 s. The PCR assay included 45 cycles of denaturation at 95 °C for 3 s and annealing/ extension at 60 °C for 20 s.

c. *In-vitro* RNA and blind panel samples. To synthesize *in vitro* RNA of Influenza A and Influenza B viruses, primers flanking the detection region were used to amplify fragments of the M gene including the region targeted by the primers and probes in the real-time PCR assays. The PCR products were cloned using the TOPO TACloning Dual Promoter Kit (Life Technologies, Burlington, Canada) and the plasmid DNA linearized using restriction enzymes (Hind III) and transcribed using the T7 RiboMAXTM Express (Promega, Madison, WI, USA). The resultant *in vitro* transcribed RNA was quantified and serial dilutions were utilized for the standard quantification process.

Validation studies were performed using a total of three blind panels: 1, A panel of six NP samples that had previously tested either positive or negative for Influenza A with a range of viral loads (crossing threshold (Ct) values ranging from 23 to 33 by qRT-PCR) (Table 1a); 2, A panel of six Influenza A positive NP and TS samples, with a range of viral loads (Crossing threshold values ranging from 24 to 32 by qRT-PCR) (Table 1b); and 3, A mixed panel of Influenza A and B positive NP specimens including a co-infected specimen (Table 1c).

Table 1. Tabular list of the three different clinical panels used to validate the performance of multiplexed assays using the fabricated micro-device.

Panel Sample No.	Sample Style	Target
(a) The Influenza A panel samples (End-point PCR)		
1	Nasopharyngeal Swab	FluA; pdm09
2	Nasopharyngeal Swab	Respiratory negative
3	Nasopharyngeal Swab	Respiratory negative
4	Nasopharyngeal Swab	FluA; pdm09
5	Nasopharyngeal Swab	Respiratory negative
6	Nasopharyngeal Swab	FluA; pdm09
(b) The Influenza A blind panel		
1	Nasopharyngeal Swab	FluA; pdm09
2	Nasopharyngeal Swab	FluA; pdm09
3	Throat Swab	FluA; pdm09
4	Nasopharyngeal Swab	FluA; pdm09
5	Nasopharyngeal Swab	FluA; pdm09
6	Nasopharyngeal Swab	FluA; pdm09
7 (+ve control)	H3 M-gene *In-vitro* RNA	FluA; H3
8 (−ve control)	PCR water	−
(c) The Influenza A, Influenza B mixed blind panel		
1	Nasopharyngeal Swab	FluA, FluB
2	Nasopharyngeal Swab	FluA, FluB
3	Nasopharyngeal Swab	FluA, FluB
4	Nasopharyngeal Swab	FluA, FluB

2.3. Experimental Procedures

A schematic diagram of the experimental set-up is shown in Figure 2a. The set-up consists of the required optical components, a microchip-PCB (Printed Circuit Board) assembly secured on a motorized *xyz* stage, an (field programmable gate array) FPGA interfaced NI PXIe-1062Q (National Instruments, Austin, TX, USA) unit for electro-actuation and feedback control, a micro-photomultiplier tube (μPMT, Hamamatsu, Japan) for continuous, scanning mode, real-time fluorescence signal read-out of the panel assays. Although it is not a packaged unit, the set-up already shows miniaturization of the multiplexed PCR unit, which is driven by the NI PXIe unit (National Instrument, Austin, TX, USA). The optical components are currently housed on a microscope platform and include: microPMT (H12400-00-01) for parallel read-out; a color charge-coupled device (CCD) camera (QImaging, Surrey, Canada) and a high speed complementary metal oxide semiconductor (CMOS) camera (Canadian Photonics Lab, Manitoba, Canada) for visual inspection and video/image capturing; a motorized *xyz* stage, controlled by an OptiScan unit (Prior Scientific) via NI program for rapid scanning and panel PCR read-outs. The operation of the resistive thermostatic zones through the NI PXIe unit has been previously described [11]. The microchip-PCB assembly (Figure 2b) utilizes a PCB (manufactured at AP Circuits, Calgary, Canada) mounted PCI ZIF test connector (Meritec Inc., Painesville, OH, USA) to secure and address the various electro-actuations and feedback controls during the multiplexed assays.

Various photomicrographs of the droplet electro-actuation based PCR thermal cycling, over the micro-device shown in Figure 1c, are illustrated in Figure 3. For all the qRT-PCR assays reported in this work, we have utilized a sealed enclosure containing PCR grade mineral oil (bioMerieux, Montreal, Canada), secured within a heated indium tin oxide (ITO)/Glass top plate, the bottom substrate and a plexiglass fixture.

The substrate was maintained at a temperature of 50 °C, required for the RT reaction which takes place on the EW electrode array, following the mixing of PCR sample and reagent droplets (see Figure 1). Furthermore to minimize thermal diffusion from the PCR droplets, the ITO/Glass top plate was also maintained at the same temperature using an isothermal plate, as shown in Figure 2b [11]. The PCR reaction volume for all the qRT-PCR assays reported in this work was kept constant at 10 μL, in order to facilitate validation studies using commercial qRT-PCR equipment at ProvLab. For each multiplexed assay, extracted RNA sample droplet (5 μL) and PCR reagent mixture droplets (5 μL) were manually pipetted and mixed using the EW electrode array, as shown in Figure 1a,b. Figure 3a shows the continuous, bi-directional actuation of a 10 μL PCR droplet following the EW based dispensing and mixing. The electrostatic/D-DEP actuation [11,23,24] is facilitated by an AC voltage (50–60 Vpp, 40 Hz), applied across a pair of herringbone electrodes upon which the droplet is electrically confined and transported. The droplet track is switched with a 50 V DC voltage applied across the top and bottom herringbone electrode pair to facilitate droplet transfer between the two temperature zones. Although the track switching is manually achieved in a timed fashion (DC bias applied after 6 s on either end of linear D-DEP actuation), it is fairly reliable due to the short track lengths and controlled droplet speed [11]. Figure 3b further illustrates the parallel thermal cycling of two identical sized (10 μL) PCR droplets, following the reverse transcription step, during a multiplexed qRT-PCR assay. The apparent increase of droplet size during the denaturing phase is expected due to the increased thermal stress that the droplet is subjected as it heats up to the higher temperature set-point. As the droplet moves out of the denaturing zone, it retains its original high contact angle, hence enabling reliable transport during multiple thermal cycles (see Figure 3b). The transport of droplets between the two thermostatic temperature zones is achieved in ~5–6 s, resulting in an effective temperature ramp rate of ~5 °C/s. The PMT read-out is carried out over the annealing zone (at 60 °C), using a linear scan of the multiple droplets, with an optical aperture set higher than the droplet diameter (twice as large as the droplet diameter) to ensure complete capture of the fluorescent signal from each droplet during the linear scan. The entire linear scan requires up to 25 s for the complete array of eight assay droplets. The captured fluorescent signal is adjusted with the background photocurrent value and plotted *vs.* PCR cycle number to obtain the complete PCR curve, reported in the results section.

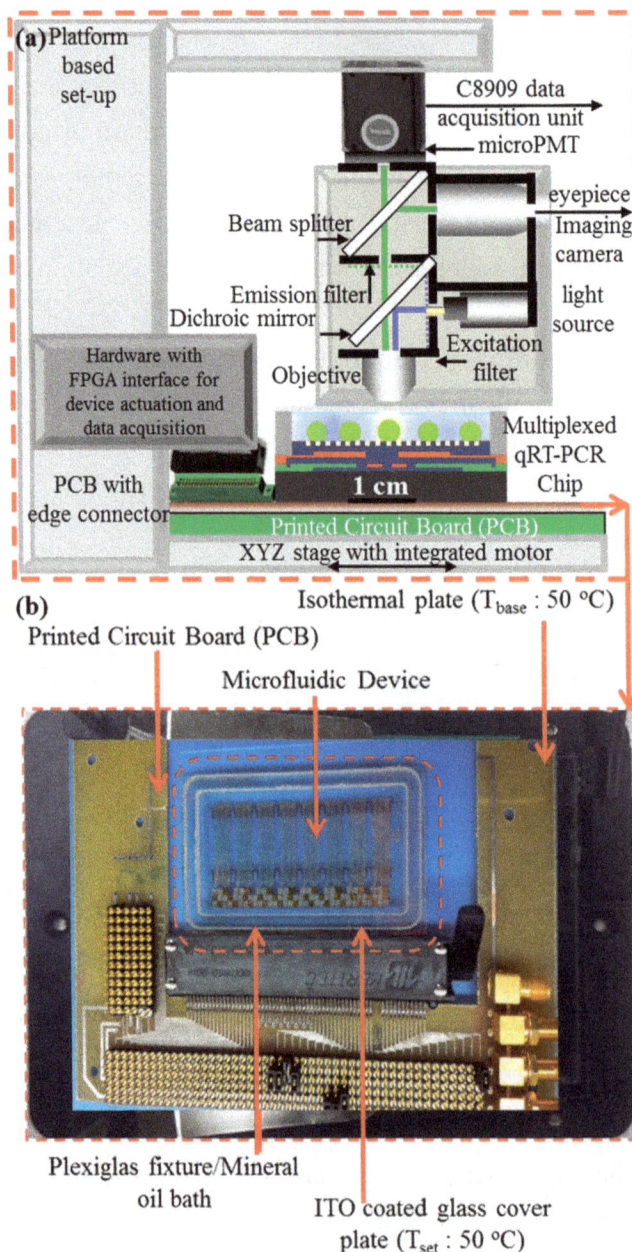

Figure 2. (a) Schematic diagram of the experimental setup; (b) An image of the microchip-PCB (Printed Circuit Board) fixture.

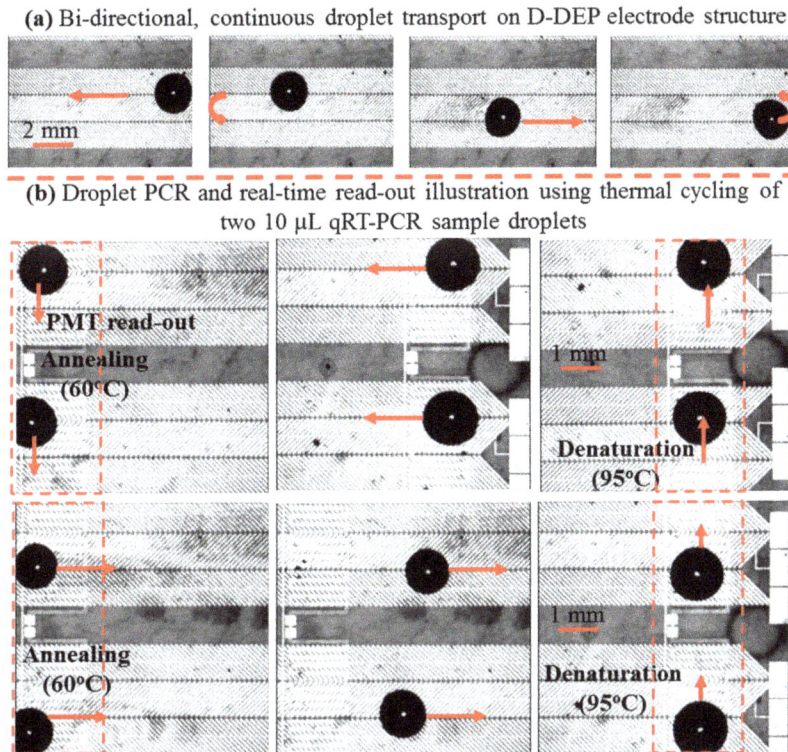

(a) Bi-directional, continuous droplet transport on D-DEP electrode structure

(b) Droplet PCR and real-time read-out illustration using thermal cycling of two 10 μL qRT-PCR sample droplets

Figure 3. Photomicrographs of **(a)** the different phases of continuous droplet transport over the newly designed bi-directional electrode scheme and **(b)** frames extracted from a real-time video showing different stages during qRT-PCR thermal cycling using two 10 μL polymerase chain reaction (PCR) droplets on a segment of the micro-device.

3. Results and Discussion

In order to validate the operation and performance of the micro-device, both end-point and quantitative RT-PCR assays were carried out on three different panels of clinically extracted patient samples (see Table 1).

3.1. Standard Quantification Curves for qRT-PCR Amplification of Spiked Influenza A and Influenza B RNA Samples

A key feature of qRT-PCR equipment is its ability to perform quantitative PCR amplification of target nucleic acid in matrix samples, with a high degree of accuracy and repeatability, over several orders of magnitude of initial template concentration. This allows the user to reliably infer the initial target DNA/RNA concentration from the qRT-PCR plots. In order to test the performance of our micro-device and furthermore to deliver quantitative outcomes on clinical samples, we used spiked *in vitro* RNA solutions, which were serially diluted and amplified simultaneously on the multiplexed array. The stock *in vitro* RNA solutions for Influenza A and B viruses were prepared as detailed in Section 2.2. The attributes of the resultant spiked samples are reported in Figure 4, which also presents the extracted qRT-PCR curves obtained for each of the 10 spiked samples (five Influenza A and five Influenza B RNA samples, shown in Figure 4). For analyzing the threshold cycle (Ct) value throughout this paper, we have set the threshold signal level based on the fluorescent noise floor of the negative control sample (see Figure 4a,b). The Ct values (averaged over two sets of multiplexed assays) were then plotted *versus* the natural log of the RNA concentration (Copy count), to report the standard quantification curves for the two target viruses (Figure 4c). The error bars, shown in the plots reported in Figure 4 and all following PCR curves, were calculated as standard deviation data from two different sets of qRT-PCR assays, conducted over two different micro-devices.

The slope (*m*) of the linear curve in Figure 4c is related to the efficiency (*E*) of the PCR as [28]:

$$E = 10^{-1/m} - 1 \qquad (1)$$

Based on Equation (1), the PCR efficiency for the Influenza A RNA samples was found to be ~95.4% whereas, the PCR efficiency for the Influenza B RNA samples was ~94.6%. The outcomes of these experiments confirmed that the designed micro-device can reliably achieve parallel and high efficiency qRT-PCR assays on multiple nucleic acid samples. Having confirmed the PCR efficiency of the micro-device, it was then used to detect the viral RNA from extracted nucleic acids from clinical samples at the ProvLab Calgary (see Section 2.2).

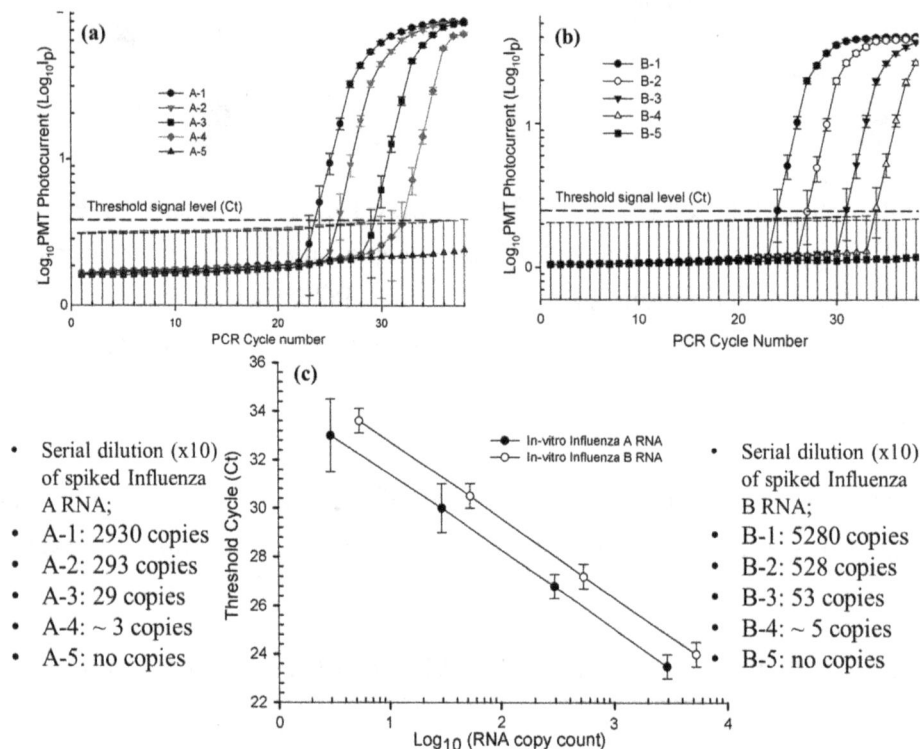

Figure 4. qRT-PCR amplification plots of (**a**) spiked Influenza A samples, (**b**) spiked Influenza B samples and (**c**) standard quantification curves for spiked Influenza A and B. samples (photocurrent, I_p in µA).

3.2. End-Point, RT-PCR Assay for a Clinical Panel of Influenza A RNA Virus

The first of the three panel assays conducted during this work used extracts from clinical samples previously characterized reported in Table 1a. A 5 µL droplet of extract from each of the six samples, along with an *in vitro* RNA sample (positive control) and a RNA free water sample (negative control) were sequentially loaded onto the respective sites (see Figure 2b) and mixed with 5 µL of PCR reagent droplets. The combined 10 µL PCR droplet was maintained at 50 °C for 5 min, for completion of the RT-reaction, before initiating parallel PCR assays.

Once the RT-reaction was complete, the eight samples were simultaneously thermally cycled for 38 PCR cycles. The motorized stage was only used at three set-points (after cycle # 10, 25 and 38) to extract the PMT photo-current read-out (see Table 2) and the PCR end-points were also recorded as CCD images, shown in Figure 5. The outcomes of this end-point parallel PCR assays, as illustrated in Figure 5 and Table 2 indicate successful identification of the eight panel samples, with the fluorescence readings and CCD images identifying samples 1, 4, 6 and 7 (+ve control) that tested positive for Influenza A virus. The three set-point PMT readings to some extent relate to the initial RNA concentration of the different panel samples as seen from Table 2.

The aberrations evident in this and other following CCD fluorescent images of PCR droplets is a result of diffraction of incident light onto locally coagulated nano-beads, which is a by-product of the soft-lithography based nano-texturing process, used during the device fabrication [25]. However, the effect of such aberrations are measured and accommodated for as the background signal levels in the PCR curves, which remain fairly constant as evident in the PCR curves reported in the following sections.

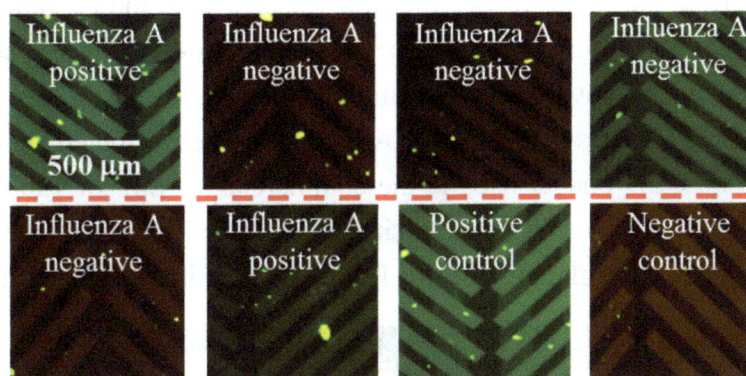

Figure 5. Charge-coupled device (CCD) images showing the outcomes (fluorescent intensity) of the end-point PCR assay carried out using panel samples of Table 1a.

Table 2. Outcomes of the end-point panel polymerase chain reaction (PCR) using samples from Table 1a.

Panel Sample No.	PMT Photocurrent at Different PCR End Points (I_p in µA)			ProvLab *Ct*
	PCR cycle # 10	PCR cycle # 25	PCR cycle # 38	
1	1.09	12.90	25.77	24
2	1.05	1.97	3.41	Negative
3	1.04	1.77	2.92	Negative
4	1.08	7.75	23.35	30
5	1.06	1.97	3.95	Negative
6	1.04	4.51	18.23	33
7 (+ve control)	1.09	15.82	30.35	29
8 (−ve control)	1.06	1.85	3.01	Negative

3.3. Quantitative, Multiplexed RT-PCR Assay on an Influenza A blind Panel

Following the successful analysis of a known panel of extracts from clinical samples using the multiplexed, end-point RT-PCR assay, we then analyzed a panel of clinical samples submitted blindly (described in Table 1b). The blind panel, prepared at ProvLab Calgary, included extracts from patients diagnosed with Influenza viral infection. The panel varied in terms of the presence/absence of the RNA virus as well as the concentration of viral load, amongst the eight samples. A positive control (sample #7) and negative control (RNA free water; sample #8) were also included in the panel. This panel was subjected to two multiplexed qRT-PCR analyses on two different micro-devices. In both analyses, the motorized stage and PMT modules were used to establish PCR curves from each of the panel samples, which are reported in Figure 6. Following the assay, the chip based PCR curves were plotted and the corresponding Ct values for each of the panel samples were analyzed and reported as the average Ct over the two micro-device based PCR assays. Subsequently, qRT-PCR reactions on the same panel of samples were also carried out at ProvLab, using the ABI 7500 Fast (Life Technologies Inc., Burlington, Canada) equipment and the Ct values from both analyses are compared in Table 3.

As is clear from Table 3 and Figure 6, the outcomes of the parallel, qRT-PCR assay using the eight panel samples on the micro-device are in agreement with the commercial PCR set-up, with accurate identification of each panel samples.

The Ct values obtained from the micro-device are in agreement with the Ct values yielded by the commercial equipment. It was noticed that the Ct values for the micro-device were consistently lower than those obtained at the ProvLab, however the variation and scalability of the two Ct value sets are almost identical. The lower Ct values for the micro-device can be attributed to a more sensitive detector (PMT compared to a CCD imager used in the commercial set-up). In Table 3, we have also reported the initial RNA copy count in each of the positively identified panel samples, estimated using the standard quantification curve for Influenza A virus RNA, reported in Section 3.1, Figure 4.

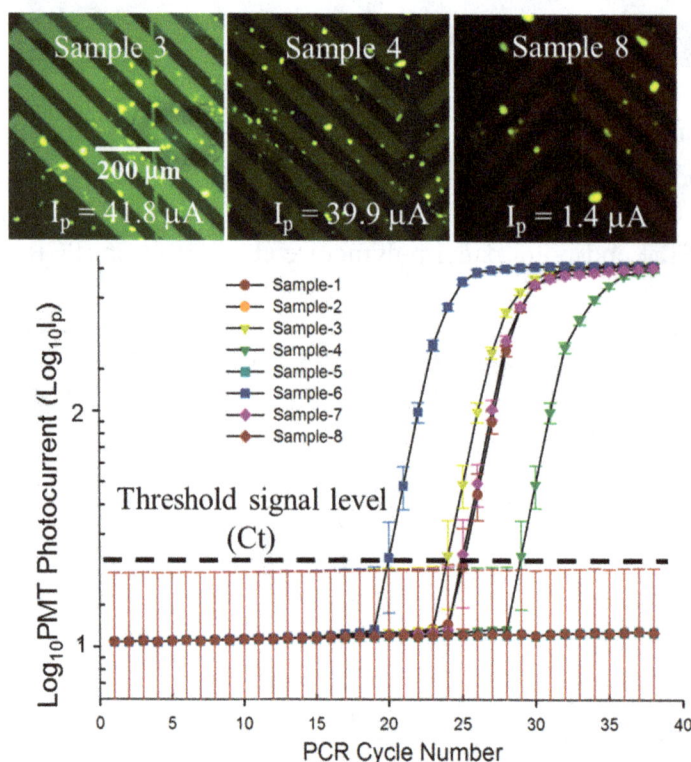

Figure 6. Plot showing qRT-PCR curves obtained during the multiplexed assay using blind panel samples of Table 1b. The fluorescent photomicrographs show a 10× magnified image, centered within the PCR droplets following 38 amplification cycles (I_p in μA).

Table 3. Outcomes of the micro-device quantitative, reverse transcription, polymerase chain reaction (qRT-PCR micro-device) assay using panel samples of Table 1b.

Panel Sample No.	Target	ProvLab *Ct*	Chip *Ct*	Initial Copies of Template RNA
1	Flu A	29	25	~590
2	Flu A	Negative	Negative	Not applicable
3	Flu A	30	26	~300
4	Flu A	32	30	~20
5	Flu A	Negative	Negative	Not applicable
6	Flu A	24	21	~3500
7 (+ve control)	Flu A	29	26	~250
8 (−ve control)	Flu A	Negative	Negative	~110

3.4. Quantitative, Multiplexed RT-PCR Assays on a Mixed, Four Sample Influenza A, Influenza B Blind Panel

The usual approach to a spectral multiplexed PCR analysis relies on the use of a multitude of primers and probes targeting each of the intended agents to be detected in the same PCR droplet. As a result of the spectral signal bandwidth and optical filtration limitations, this results in practice in limiting the multiplexing capabilities to up to five to six targets per PCR assay. The development of our micro-device was inspired by the notion of incorporating both spectral and spatial multiplexing, where multiple targets can be amplified and read-out in a parallel and automated fashion.

In order to demonstrate this versatile multiple sample target handling, we investigated a mixed blind panel of clinical samples, as shown in Table 1c, which contained different initial concentrations of Influenza A and Influenza B viral RNA, prepared from patient samples extracted at ProvLab Calgary. The synthesized molecular probes for the two RNA targets were labeled respectively with FAM™ (λ_{ex}./λ_{em}.: 492 nm/520 nm) and VIC™ (λ_{ex}./λ_{em}.: 538 nm/554 nm) fluorophores. The four panel samples from Table 1c were then paired in binary combination with the reagent mix droplets containing one of the two fluorescent markers and transported to the eight droplet tracks.

The eight 10 µL PCR droplets were then amplified over 38 PCR cycles and analyzed during the annealing phase of each cycle, through the continuous mode PMT read-out. The multiplexed assays (38 PCR cycles and RT reaction), which were repeated on two different micro-devices, were completed within 40 min from sample/reagent loading onto the micro-device to the determination of all qRT-PCR curves (and the corresponding *Ct* values). The extracted data was plotted and the resulting qRT-PCR curves are reported in Figure 7. After completion of the thermal cycling, fluorescent CCD images were captured showing the eight PCR droplets (see Figure 7). It is clear from the CCD fluorescent images, and from the curves, that sample 1 tested positive for Influenza A virus, sample 2 tested positive for both Influenza A and Influenza B viruses, sample 3 tested negative for both RNA viruses and sample 4 tested positive for Influenza B virus. The *Ct* values, analyzed from the PMT data and averaged over two different micro-device based qRT-PCR assays, are reported in Table 4, alongside the *Ct* values measured with the ABI 7500 fast, at ProvLab Calgary and an estimated initial RNA template copy number. Clearly the multiplexed assay on the micro-device successfully analyzes the mixed blind panel of Influenza A and B viruses and accurately reflected their relative concentrations.

These findings support our contention that a combination of spatial and spectral multiplexing will significantly extend the current limitations of the conventional multiplexed qRT-PCR methodology.

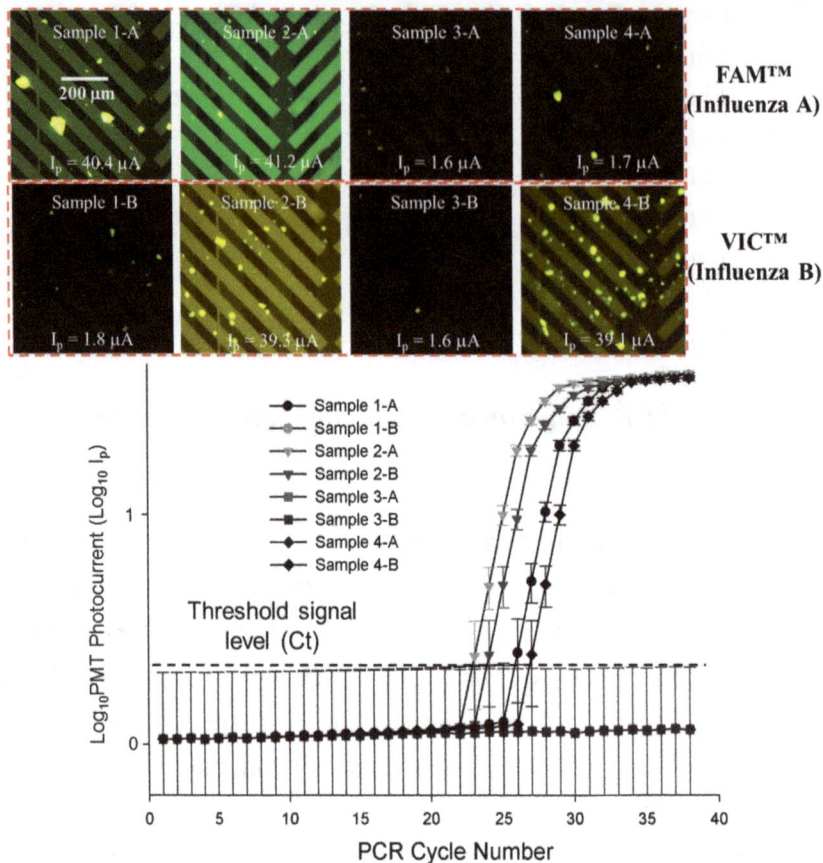

Figure 7. Photomicrographs showing the fluorescent images corresponding to the eight PCR droplets and the extracted plot of the eight qRT-PCR curves (I_p in μA).

Table 4. Outcomes of the micro-device qRT-PCR assay using mixed panel samples of Table 1c.

Panel Sample No.	Target	ProvLab *Ct*	Chip *Ct*	Initial Copies of Template RNA
1-A	Flu A	29	27	~290
1-B	Flu B	Negative	Negative	Not applicable
2-A	Flu A	27	24	~2900
2-B	Flu B	28	25	~1050
3-A	Flu A	Negative	Negative	Not applicable
3-B	Flu B	Negative	Negative	Not applicable
4-A	Flu A	Negative	Negative	Not applicable
4-B	Flu B	30	28	~110

4. Conclusions

This present investigation demonstrates and furthermore extends the applicability of the continuous D-DEP based droplet transport method for parallel, spatially multiplexed qRT-PCR reactions on a nano-textured DMF chip. The improved micro-electrode architecture accommodates up to eight parallel, qRT-PCR reactions. As a proof of principle, detection of Influenza A and B viruses from clinical samples was conducted using a blind panel. Influenza A and B were accurately identified and quantified using the standard quantification method, in the two micro-device based qRT-PCR assays. The outcomes of the repeated blind panel experiments confirm that the micro-device can successfully handle more than one nucleic acid samples and markers over an array of parallel, spatially multiplexed DMF micro-electrodes, to screen for a panel of viral/infectious diseases. The efficiency of chip based qRT-PCR assays were reasonably within the accepted industrial benchmark (PCR efficiency ~94%–97%) and the completion time for the sample loading/mixing, RT-reactions and up to 38 PCR thermal cycles for up to eight different PCR droplets was found to be ~35–40 min, again comparable to that of a commercial fast qRT-PCR equipment. The detection limit, as identified using the chip based standard quantification process, for the multiplexed qRT-PCR micro-device was found to be <10 copies of RNA templates/PCR reaction. The micro-device furthermore offers future integration of both spatial (parallel qPCR reactions with differed targets) and spectral (multiple target markers in same PCR assay) multiplexing to screen for a larger panel of infectious agents. As a next step in the development, our focus is to improve the up-stream sample handling to achieve serial dilution of RNA samples and facilitate on-chip mixing and preparation of the reagent mixture and dispensing of multitude of sample droplets to suitably address the multiplexed qRT-PCR tracks. In addition, we will focus on the development of a separate sample extraction and purification chip to separate, lyse and concentrate target DNA/RNA from clinical patient samples, in preparation for the qRT-PCR amplification and detection stage. These proposed developments will lead to a portable sample-to-detection microsystem, suitable for example for field analysis of human, live-stock and food borne pathogens.

Acknowledgments

The authors gratefully acknowledge the financial support received from National Science and Engineering Research Council of Canada (NSERC) under the Discovery Grant program and that provided by CMC Microsystems (Canada) to support the microfabrication of the devices utilized in the study. The authors are furthermore thankful to the Provincial Laboratory for Public Health of Alberta for providing the extracted clinical samples and PCR reagents used in this work.

Author Contributions

Ravi Prakash designed and fabricated the DMF micro-chips, conducted the on-chip, qRT-PCR experiments, data analysis and contributed to the manuscript writing and revision process; Kanti Pabbaraju and Sallene Wong contributed in sample extraction, off-chip validation experiments and manuscript writing; Anita Wong prepared the extracted clinical samples and PCR reagents for both off-chip and on-chip qRT-PCR experiments and validated the micro-chip experimental data; Raymond Tellier contributed to the planning of the validation experiments and assisted in paper writing and revision; Karan Kaler contributions include the design of the developed DMF micro-chips; facilitated the DMF micro-chip based qRT-PCR experiments and contributed to the writing, revision and proofreading of the manuscript.

Conflicts of Interest

The authors declare no conflict of interest.

References

1. Saiki, D.H.; Gelfand, S.; Stoffel, S.; Scharf, S.J.; Higuchi, R.; Horn, G.T.; Mullis, K.B.; Erlich, H.A. Primer-directed enzymatic amplification of DNA with a thermostable DNA polymerase. *Science* **1988**, *239*, 487–491.

2. Langin, T.; Robert, T. Mullis, K. B., Ferré, F. and Gibbs, R. A. (Eds.). 1994. The polymerase chain reaction (PCR). Birkhäuser Verlag AG, Basel, Switzerland. ISBN 3-7643-3607-2 (H.b.). ISBN 3-7643-3750-8 (P.b.). *J. Evol. Biol.* **1994**, *8*, 399–401.

3. Markoulatos, P.; Siafakas, N.; Moncany, M. Multiplex polymerase chain reaction: A practical approach. *J. Clin. Lab. Anal.* **2002**, *16*, 47–51.

4. Logan, J.; Edwards, K.J.; Saunders, N. *Real-Time PCR: Current Technology and Applications*; Caister Academic Press: Norfolk, UK, 2009.

5. Wittmer, C.T.; Kusukawa, N. Real-time PCR and melting analysis. In *Molecular Microbiology Diagnostic Principles and Practice*, 2nd ed.; ASM Press: Washington, DC, USA, 2011.

6. Kopp, M.U.; de Mello, A.J.; Manz, A. Chemical amplification: Continuous-flow PCR on a chip. *Science* **1998**, *280*, 1046–1048.

7. Manz, A.; Becker, H. *Microsystem Technology in Chemistry and Life Sciences*; Springer: Berlin, Germany, 1998.

8. Lagally, E.T.; Emrich, C.A.; Mathies, R.A. Fully integrated PCR-capillary electrophoresis microsystem for DNA analysis. *Lab Chip* **2001**, *1–2*, 102–107.

9. Liu, R.H.; Yang, J.N.; Lenigk, R.; Bonanno, J.; Grodzinski, P. Self-contained, fully integrated biochip for sample preparation, polymerase chain reaction amplification, and DNA microarray detection. *J. Anal. Chem.* **2004**, *76*, 1824–1831.

10. Tewhey, R.; Warner, J.B.; Nakano, M.; Libby, B.; Medkova, M.; David, P.H.; Kotsopoulos, S.K.; Samuels, M.L.; Hutchison, J.B.; Larson, J.W.; Topol, E.J.; Weiner, M.P.; Harismendy, O.; Olson, J.; Link, D.R.; Frazer, K.A. Microdroplet-based PCR enrichment for large-scale targeted sequencing. *Nat. Biotechnol.* **2009**, *27*, 1025–1031.

11. Prakash, R.; Pabbaraju, K.; Wong, S.; Wong, A.; Tellier, R.; Kaler, K.V.I.S. Droplet microfluidic chip based nucleic acid amplification and real-time detection of influenza viruses. *J. Electrochem. Soc.* **2014**, *161*, 3083–3093.

12. Dahl, A.; Sultan, M.; Jung, A.; Schwartz, R.; Lange, M.; Steinwand, M.; Livak, K.; Lehrach, H.; Nyarsik, L. Quantitative PCR based expression analysis on a nanoliter scale using polymer nano-well chips. *Biomed. Microdev.* **2007**, *9*, 307–314.

13. Freire, V.S.; Ebert, A.D.; Kalisky, T.; Quake, S.R.; Wu, J.C. Microfluidic single-cell real-time PCR for comparative analysis of gene expression patterns. *Nat. Protoc.* **2012**, *7*, 829–838.

14. Saunders, D.C.; Holst, G.L.; Phaneuf, C.R.; Pak, N.; Marchese, M.; Sondej, N.; McKinnon, M.; Forest, C.R. Rapid, quantitative, reverse transcription PCR in a polymer microfluidic chip. *Biosens. Bioelec.* **2013**, *44*, 222–228.

15. Lee, S.H.; Kim, S.W.; Kang, J.Y.; Ahn, C.H. A polymer lab-on-a-chip for reverse transcription (RT)-PCR based point-of-care clinical diagnostics. *Lab Chip* **2008**, *8*, 2121–2127.

16. Roberts, C.H.; Jiang, W.; Jayaraman, J.; Trowsdale, J.; Holland, M.J.; Traherne, J.A. Killer-cell Immunoglobulin-like receptor gene linkage and copy number variation analysis by droplet digital PCR. *Genome Med.* **2014**, *6*, 1–9.

17. Pollack, M.G.; Fair, R.B.; Shenderov, A.D. Electrowetting-based actuation of liquid droplets for microfluidic applications. *Appl. Phys. Lett.* **2000**, *77*, 1725–1726.

18. Cho, S.K.; Moon, H.; Kim, C.J. Creating, Transporting, Cutting, and Merging Liquid Droplets by Electrowetting-Based Actuation for Digital Microfluidic Circuits. *J. Microelectromech. Syst.* **2003**, *12*, 70–80.

19. Sista, R.; Hua, Z.; Thwar, P.; Sudarsan, A.; Srinivasan, V.; Eckhardt, A.E.; Pollock, M.; Pamula, V. Development of a digital microfluidic platform for point of care testing. *Lab Chip* **2008**, *8*, 2091–2104.

20. Hua, Z.; Rouse, J.L.; Eckhardt, A.E.; Srinivasan, V.; Pamula, V.K.; Schell, W.A.; Benton, J.L.; Mitchell, T.G.; Pollack, M.G. Multiplexed Real-Time Polymerase Chain Reaction on a Digital Microfluidic Platform. *Anal. Chem.* **2010**, *82*, 2310–2316.

21. Srinivasan, V.; Pamula, V.K.; Fair, R.B. An integrated digital microfluidic lab-on-a-chip for clinical diagnostics on human physiological fluids. *Lab Chip* **2004**, *4*, 310–315.

22. Chang, Y.H.; Lee, G.B.; Huang, F.C.; Chen, Y.Y.; Lin, J.L. Integrated polymerase chain reaction chips utilizing digital microfluidics. *Biomed. Microdev.* **2006**, *8*, 215–225.

23. Gunji, M.; Nakanishi, H.; Washizu, M. Droplet actuation based on single-phase electrostatic excitation. In Proceedings of 8th International Conf. on Miniaturized Systems in Chemistry and Life Sciences (μ-TAS2004), Malmö, Sweden, September 26–30 2004; Volume 1, pp. 168–170.

24. Kaler, K.V.I.S.; Prakash, R.; Chugh, D. Liquid Dieletrophoresis and Surface Microfluidics. *Biomicrofluidics* **2010**, *4*, 022805.

25. Prakash, R.; Papageorgiou, D.P.; Papathanasiou, A.G.; Kaler, K.V.I.S. Dielectrophoretic liquid actuation on nano-textured super hydrophobic surfaces. *Sens. Actuators B Chem.* **2013**, *182*, 351–361.

26. Lagally, E.T.; Simpson, P.C.; Mathies, R.A. Monolithic integrated microfluidic DNA amplification and capillary electrophoresis analysis system. *Sens. Actuators B Chem.* **2006**, *63*, 138–146.

27. Zhong, R.; Pan, X.; Jiang, L.; Dai, Z.; Qin, J.; Lin, B. Simply and reliably integrating micro heaters/sensors in a monolithic PCR-CE microfluidic genetic analysis system. *Electrophoresis* **2009**, *30*, 1297–1305.

28. Pfaffl, M.W. A new mathematical model for relative quantification in real-time RT-PCR. *Nucleic Acids Res.* **2001**, *29*, 2002–2007.

8

Predictable Duty Cycle Modulation through Coupled Pairing of Syringes with Microfluidic Oscillators

Sasha Cai Lesher-Perez [1,†], Priyan Weerappuli [1,2,3,†], Sung-Jin Kim [1,4], Chao Zhang [1,5,6] and Shuichi Takayama [1,7,8,*]

[1] Department of Biomedical Engineering, Biointerfaces Institute, University of Michigan, Ann Arbor, MI 48109, USA; E-Mails: sashacai@umich.edu (S.C.L-P.); pweerapp@umich.edu (P.W.); yahokim@konkuk.ac.kr (S.-J.K.); sclzzc@gmail.com (C.Z.)
[2] Department of Biomedical Engineering, Wayne State University, Detroit, MI 48202, USA
[3] Department of Physiology, Wayne State University, Detroit, MI 48201, USA
[4] Department of Mechanical Engineering, Konkuk University, Seoul 143-701, Korea
[5] Key Laboratory of Low-Grade Energy Utilization Technologies and Systems, Chongqing University, Chongqing 400030, China
[6] Institute of Engineering Thermophysics, Chongqing University, Chongqing 400030, China
[7] Department of Macromolecular Science and Engineering, University of Michigan, Ann Arbor, MI 48109, USA
[8] Division of Nano-Bio and Chemical Engineering World Class University Project, Ulsan National Institute of Science and Technology, Ulsan 689-798, Korea

[†] These authors contributed equally to this work.

[*] Author to whom correspondence should be addressed; E-Mail: takayama@umich.edu

External Editor: Jeong-Bong Lee

Abstract: The ability to elicit distinct duty cycles from the same self-regulating microfluidic oscillator device would greatly enhance the versatility of this micro-machine as a tool, capable of recapitulating *in vitro* the diverse oscillatory processes that occur within natural systems. We report a novel approach to realize this using the coordinated modulation of input volumetric flow rate ratio and fluidic capacitance ratio. The demonstration uses a straightforward experimental system where fluid inflow to the oscillator is provided by two syringes (of symmetric or asymmetric cross-sectional area) mounted upon a single syringe pump applying pressure across both syringes at a constant linear velocity. This produces distinct volumetric outflow rates from each syringe that are proportional to the ratio between their cross-sectional areas. The difference in syringe cross-sectional area also leads to differences in fluidic capacitance; this underappreciated capacitive difference allows us to present a simplified expression to determine the microfluidic oscillators duty cycle as a function of cross-sectional area. Examination of multiple total volumetric inflows under asymmetric inflow rates yielded predictable and robust duty cycles ranging from 50% to 90%. A method for estimating the outflow duration for each inflow under applied flow rate ratios is provided to better facilitate the utilization of this system in experimental protocols requiring specific stimulation and rest intervals.

Keywords: microfluidics; microfluidic oscillator; duty cycle

1. Introduction

Emerging interest in microfluidic machines that directly utilize fluidic energy to execute core operations has prompted the development of self-regulated machines that, by virtue of their autonomous operation, have also garnered much attention as potential platforms for basic biomedical research [1–3].

Biological and physiological systems are fundamentally regulated by oscillatory processes operating at discrete spatial and temporal scales. Our understanding of these systems, consequently, has benefited from the development of pulsatile stimulation techniques capable of manipulating the temporal dynamics of these processes and investigating the role of timing within them. Historically, the *in vitro* study of these processes in cultured cells was advanced primarily by two types of assays: one in which a single stimulus is bath-applied and later washed off (e.g., pulse-chase analysis [4,5], and BrdU "birth dating" [6]); and one in which a continuous long-term temporal stimulation pattern is applied by way of an external control apparatus [7]. Advancements in microfluidic technology have catalyzed the translation of such assays, in parallel with the development of novel counterparts, to forms supported by these emerging micro-scale—"lab-on-a-chip"—platforms [8–11].

Microfluidic devices often emulate electronic circuitry and utilize integrated conduits and embedded valves to direct and manipulate fluid flows. The control systems underlying their operation, however, have typically remained external from the fluidic devices themselves [12–14]. An awareness that this rise in peripheral equipment cost may limit "next-generation" microfluidic systems has motivated the development of autonomous, pre-programmed, fluidic systems [1,12–17]. Foremost among these is the microfluidic oscillator [18].

Not unlike how electronic oscillators were among the first broadly adopted automated electrical circuits; self-oscillating microfluidic devices provide a simple, yet useful, first target for microfluidic automation [1,2] as evidenced by the growing body of literature describing experimental methods, wherein cells cultured within micro-devices are chemically stimulated in a pulsatile, rather than continuous, manner [8,10,19,20]. One such method for cellular interrogation modifies stimulation events by altering the duration of an applied stimulus and/or rest period; effectively manipulating the oscillation frequency and duty cycle of the stimulatory system [19]. Through this approach, it has been observed that different responses may be elicited from the same population of cells by manipulating these stimulatory parameters.

The work presented here was motivated by the questions: how can a single microfluidic oscillator circuit be designed to best support multiple stimulatory frequencies and rest periods; and how can this be done in a manner that is easy to understand and perform by non-microfluidic experts? We have previously demonstrated the ability to alter oscillation frequency by modifying flow rate, and to alter duty cycle by modifying the device itself [1,2]. As the technical burden of repeatedly designing and fabricating different devices for each desired duty cycle is both difficult and tedious; we asked if a continuous and predictable modification of duty cycle could be achieved by simply modifying the syringes used to provide volumetric inflow.

The challenge associated with modifying volumetric inflow rate lies in the effect this may have upon the threshold opening pressure of each valve [21]. Due to the complexity of the relationship between volumetric flow rate and duty cycle, predicting the duty cycle resulting from a change in volumetric flow rate is not trivial. Additional challenges arise if two syringe pumps are used to generate differing volumetric inflows, owing largely to inherent pump-to-pump variability and general inflow rate unsteadiness that may produce unstable oscillations [22]. Here we report the predictable modulation of duty cycle using two syringes mounted upon a single syringe pump such that volumetric flow rate ratio and fluidic capacitance are coupled. This setup is advantageous in that it allows duty cycle to be considered simply as a function of the volumetric inflow rate ratio; requiring no modifications of the microfluidic circuit to robustly produce distinct duty cycles.

2. Working Principle

The microfluidic oscillator functions by converting two constant volumetric flow rate inflows to one oscillatory outflow through the activity of two normally-closed three-way valves that generate oscillations in fluid outflow through the alternate obstruction of each inflow (Figure 1).

Briefly, if we denote the two valves *valve 1* and *valve 2*, and arbitrarily assume that *valve 2* is initially in an open position—allowing fluid to flow across it; a portion of the outflow from *valve 2* will be diverted from its drain terminal to the gate terminal of *valve 1*. The gate terminal refers to the conduit leading to the region below the membrane valve unit. The accumulation of fluid within this region supplies the gate pressure of *valve 1* (P_{G1}); preventing the downward deflection of the membrane, and consequently preventing *valve 1* from transitioning to an open position while P_{G1} exceeds the source pressure of *valve 1* (P_{S1}) generated by the accumulation of fluid in the portion of the valve upstream from the *valve 1* gate.

When P_{S1} has surpassed the sum of P_{G1} and the inherent pressure threshold of *valve 1* (P_{th1}), determined by the specific mechanical properties of the membrane, the membrane is deflected downward, and fluid is allowed to travel through *valve 1*. A portion of this outflow is then diverted from its drain terminal to the gate terminal of *valve 2*, as the outflow from *valve 1* had been diverted previously, and supplies the gate pressure necessary to force the accumulation of fluid upstream of *valve 2*, until the difference between P_{S2} and P_{G2} has exceeded P_{th2} (Figure 1a,b). The coordination of these processes, resulting in the anti-synchronized opening and closing of both valve units, produces an oscillatory outflow (described in greater detail in previous work [2,21]).

Figure 1. Schematic for the experimental system. The three panels displayed represent the behavior of the microfluidic oscillator at three time points during operation under symmetric flow conditions. (**a**) Two fluids (blue and red) are introduced through two syringes mounted on a single syringe pump. The fluids enter the device at a constant rate, but are converted into an oscillatory outflow when passing through the valves. (**b**) A cross section of each valve unit at the time points displayed in panel (**a**). Initially, the source pressure (P_{S1}) is insufficient ($P_{S1} < P_{G1} + P_{th1}$) to displace the membrane downward, allowing the blue fluid to outflow. When the pressure has reached its maximum value (P_{max}), the membrane is displaced ($P_{S1} > P_{G1} + P_{th1}$), allowing the red fluid to outflow until sufficient source pressure (P_{S2}) has accumulated within the chamber above the opposite membrane ($P_{S2} > P_{G2} + P_{th2}$) allowing the blue fluid to outflow. (**c**) The time points within the pressure data time series corresponding to the valve and outflow profiles presented in panels (**a**) and (**b**) are indicated. A sample P_{max} and P_{th} are also represented, as well as the relationship between inflow rate (Q_i), internal capacitance (C_i), and external capacitance (C_e).

[a]

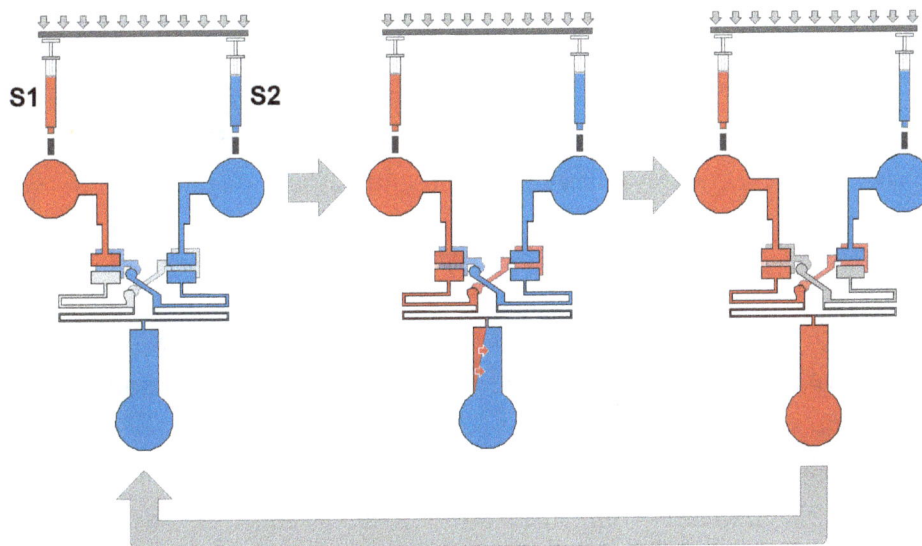

[b]

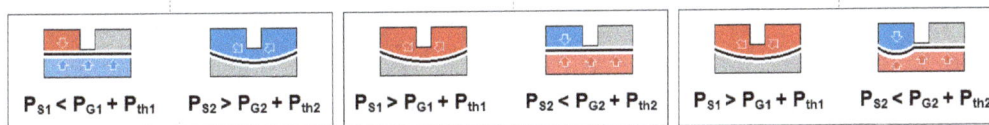

[c]

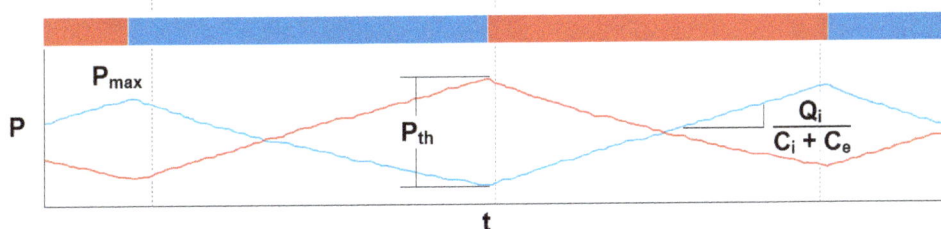

Functionally, as the gate pressure of the valve regulating one flow is itself regulated by the volumetric outflow rate across the other, we assume the following characteristic:

$$Q_{in} = C \times \frac{dP}{dt} \tag{1}$$

$$Q_{in} = C \times \frac{P_{th}}{T_{off}} \tag{2}$$

This expression, where Q_{in}, C and P represent inflow rate, fluidic capacitance, and pressure respectively, may be expanded to describe the threshold-dependent mechanism underlying the functionality of the valves. Conceptually, the transition between a closed-to-open or open-to-closed valve-state is governed by the values of P_{th} and P_G set by the mechanical properties of the membrane and buildup of fluid pressure below the membrane (Figure 1b), respectively, and the rate at which fluid pressure builds within the valve region above the membrane (P_S) [2]. The relationship between inflow rate and capacitance, thus, may be used to determine duty cycle as a function of time:

$$\frac{T_1}{T_1 + T_2} = \frac{\dfrac{C_1 \times P_{th1}}{Q_1}}{\left[P_{th1} \times \dfrac{C_1}{Q_1}\right] + \left[P_{th2} \times \dfrac{C_2}{Q_2}\right]} \tag{3}$$

Under symmetric flow conditions, $Q_1 \cong Q_2$, where the mechanical properties of the membrane and valve compartments are preserved across both valves, the assumption is $P_{th1} \cong P_{th2}$ and $C_1 \cong C_2$, allowing us to consequently define duty cycle solely as a function of volumetric flow rate.

$$\frac{T_1}{T_1 + T_2} \approx \frac{Q_1 + Q_2}{Q_1} \tag{4}$$

Equation (4) depicts an attractive relationship that relates duty cycles simply to volumetric inflow ratios. By this definition, the introduction of asymmetry to the volumetric inflow rates of each fluid, Q_i, would produce asymmetric duty cycles. However, in asymmetric conditions where $Q_1 \neq Q_2$ (e.g., $Q_1 < Q_2$), the syringe supplying the greater volumetric inflow (Q_2) will result in a greater threshold pressure for the valve regulating the lesser volumetric inflow, and consequently, $P_{th1} > P_{th2}$. The presence of this asymmetry suggests that the use of two identical syringes, evacuated at asymmetric linear velocities, would rely upon a complex balance between Q_{in}, C, and P such that the duty cycles produced may not be accurately modeled by Equation (4). One way to maintain the relationship shown in Equation (4) would be to modulate C_i together with Q_i so that $P_{thi} \times C_i \approx$ constant. One way to achieve this conveniently is by mounting two plastic syringes of different cross-sectional area on one syringe pump (Figure 2), and utilizing the compliance of the syringe components [23] and resulting capacitive differences of the syringes [12]. Within the described system, as syringe outflow rate is a function of velocity and syringe cross-sectional area, and as both syringes are evacuated at the same linear velocity, we may further refine our definition of duty cycle as being a function of syringe diameter (Figure 2b). By using syringes of different diameters, we apply Equation (4) and demonstrate predictability of duty cycle values as a function of the combination of syringes used (Table 1).

Figure 2. Schematic for the experimental generation of symmetric and asymmetric volumetric flow rates, and changes in duty cycle and pressure profile produced as a function of syringe diameter. (**a**) Two sample conditions where *Syringe 1*, (red), and *Syringe 2*, (blue), are mounted on a single syringe pump. The ratios illustrated are the symmetric 3 mL:3 mL (upper) and asymmetric 3 mL:60 mL (lower). Within the experimental protocol , *Syringe 1* was held constant in all pairings while *Syringe 2* was varied to achieve symmetric (50%) and asymmetric (>50%) duty cycles; and total volumetric inflow rate remained constant. Experimentally generated pressure profile waveforms are presented against alternating background bands representing the fluid outflow profile. (**b**) Pressure profile and stimulation period for the four inflow ratio regimes. Pressure profiles were generated while the syringe pump was moving at a constant linear velocity such that the total volumetric inflow rate (the sum of the inflows supplied by each syringe) was maintained at a volumetric flow rate of 20 μL/min. The pressure profiles recorded (*P*) are presented above each trace representing the concentration of a fluidic stimulant ([*S*]) provided via *Syringe 2*, in the outflow.

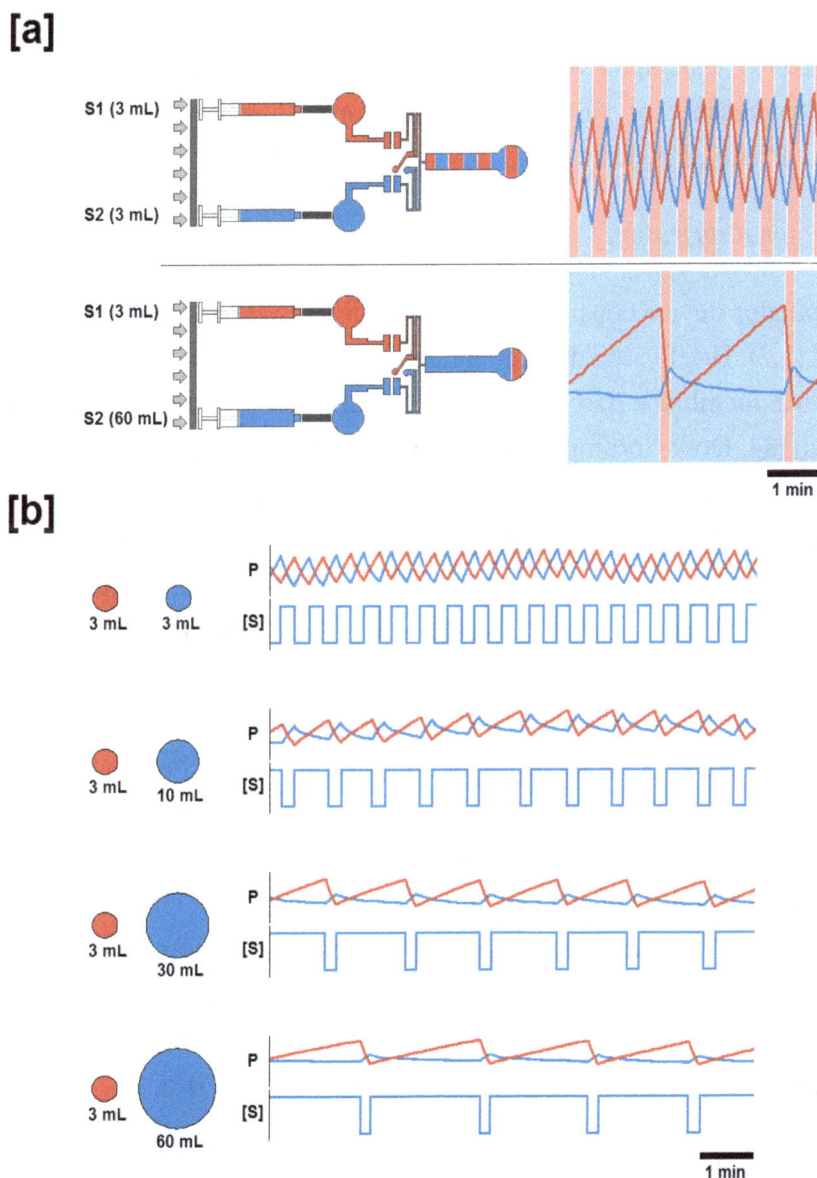

3. Materials and Methods

3.1. Master Mold Fabrication

Microfluidic oscillator master molds were fabricated upon 4″-silicon wafers using the negative photoresist, SU-8 (MicroChem, Newton, MA, USA). Following air-cleaning of the wafer, SU-8 2075 photoresist was deposited on the wafer and spin-coated at 500 rpm (acceleration of 440 rpm/s) for 10 s and at 2100 rpm (acceleration of 440 rpm/s) for 30 s. The coated wafer was then placed on a hotplate for pre-exposure baking at 65 °C for 5 min, 95 °C for 20 min and then allowed to gradually cool to room temperature by allowing it to remain on the hotplate after the plate was turned off. The SU-8 substrate was then exposed with conventional UV (~17 mJ/cm²) for 30 s using a mask aligner (Hybrid Technology Group), and then placed on a hotplate for post-exposure baking at 65 °C for 5 min, 95 °C for 10 min and then allowed to gradually cool to room temperature as before. Unexposed regions of photoresist were dissolved by repeatedly immersing the wafer in fresh SU-8 developer solution (MicroChem, Newton, MA, USA) for 60 s intervals until all non-exposed/cross-linked regions of SU-8 were removed. The completed mold was then placed within a gravity convection oven (DX-400, Yamato Scientific America, Santa Barbara, CA, USA) for 15 min at 120 °C and, upon returning to room temperature, was treated (silanized) in a desiccator for 1 h in the presence of vaporized tridecafluoro-1,1,2,2-tetrahydrooctyl-1-trichlorosilane (United Chemical Tech., Bristol, PA, USA).

3.2. Microfluidic Oscillator Fabrication

The microfluidic oscillator device consists of three polydimethylsiloxane (PDMS) layers assembled as previously described [1,2]. Briefly, the device features (100 μm height) were imprinted in the top and bottom layers, and a PDMS membrane (target thickness: 20 μm) was positioned between them (Figure 1).

1:10 PDMS (Sylgard 184, Dow Corning, Midland, MI, USA) was poured onto the master mold and allowed to cure within a gravity convection oven at 60 °C for 6 h. The cured PDMS slab was then removed from the mold and cut into individual device layers. Concurrently, PDMS membranes were fabricated by spin-coating 1:10 PDMS onto glass slides pre-treated with silane as before. PDMS membranes were then cured within a gravity convection oven for 5 min at 120 °C and 10 min at 60 °C. Prior to final assembly, a 2-mm biopsy punch was used to remove PDMS from the inlet and outlet ports of the top device layer. The bottom layer and membrane were then treated by plasma oxidation (Covance MP, FemtoScience, Hwaseong-si, Gyeonggi-do, South Korea) to facilitate bonding and, following bonding, were then placed in a gravity convection oven at 120 °C for 5 min and at 60 °C for 10 min. Thru-holes were then made in the membrane to allow fluid communication between the top and bottom device layers, using a 350-μm biopsy punch (Ted Pella Inc., Redding, CA, USA). The top layer was then treated by plasma oxidation to facilitate bonding with the membrane-bottom layer assembly. Following treatment, but preceding bonding, the normally closed region of the top layer was "deactivated" by being brought into direct contact with an unoxidized PDMS "stamp". Following final bonding, assembled devices were incubated for 2 min within a gravity convection oven at 120 °C.

3.3. Microfluidic Oscillator Testing and Data Processing

Microfluidic oscillators were tested by connecting pressure sensors (Model 142PC05D, Honeywell, NJ, USA) at the device inlets via Tygon tubing (Saint-Gobain™ Tygon™ R-3603 Clear Laboratory Tubing, Saint-Gobain Performance Plastics, Akron, OH, USA) to measure source pressure. Source pressure data was collected for both valves to quantify pressure buildup and release corresponding to fluid accumulation and evacuation, respectively, through the valves; our previous work highlighted the relationship between source pressure and drain pressure [24]. The occurrence of fluidic oscillations and the coincident timing of these oscillations relative to source pressure profiles were verified visually. All subsequent quantification and assessment, however, was performed using source pressure data. Data was obtained at a sampling rate of 1000 Hz, every 100 data points were averaged (resulting in 1 data point per 100 ms), and stored using LabVIEW (National Instruments, Austin, TX, USA). Data was recorded for a minimum of four hours, of which the data acquired during the first hour for each condition was examined and discarded to ensure the volumetric flow and capacitance of the fluidic system had stabilized, and only the subsequent time (three hours) was assessed. Syringe pumps (Model KDS220, KD Scientific, Holliston, MA, USA and Model Fusion 200, Chemyx, Stafford, TX, USA) were used to provide constant volumetric flow to the device. One input, a 3 mL syringe (*Syringe 1*) remained connected to one inlet port for the entirety of the study, while the second (*Syringe 2*) was allowed to alternate between 3 mL, 10 mL, 30 mL and 60 mL plastic syringes (Becton, Dickinson and Company, Franklin Lakes, NJ, USA). The syringe pump was programmed with total volumetric inflow rates appropriate for each syringe pairing, such that $Q_2 \geq Q_1$ and $Q_2 + Q_1 = Q_{total}$.

Voltage data were collected using LabVIEW and processed, in part, using the open-source peakdet [25].

4. Results and Discussion

4.1. Predictive Duty Cycle Control

Using Equation (4), we calculated and experimentally measured duty cycle as a function of volumetric flow rate ratios achieved through the simple utilization of two plastic syringes of different cross-sectional area mounted on a single syringe pump (Table 1). The estimates generated by Equation (4) agreed with experimental observations.

Table 1. Different syringe pairings on a single syringe pump enables different duty cycles to be achieved, while maintaining a constant total volumetric inflow rate of 20 µL/min.

Syringe 1			*Syringe 2*			Duty Cycle (Expected)	Duty Cycle (Observed)
Volume (mL)	Diameter (mm)	Inflow Rate (µL/min)	Volume (mL)	Diameter (mm)	Inflow Rate (µL/min)		
3	8.66	15.33	1	4.78	4.67	23.35%	-
3	8.66	10.00	3	8.66	10.00	50.00%	48.71%
3	8.66	6.80	5	12.06	13.20	65.98%	-
3	8.66	5.26	10	14.5	14.74	73.71%	74.59%
3	8.66	3.40	20	19.13	16.60	82.99%	-
3	8.66	2.75	30	21.7	17.25	86.26%	86.00%
3	8.66	1.90	60	26.7	18.10	90.48%	90.82%
3	8.66	0.97	140	38.4	19.03	95.16%	-

Highlighted values represent syringe combinations studied experimentally. The duty cycle values presented are calculated with respect to *Syringe 2*. Utilizing this system, we succeeded in achieving duty cycles ranging from 50% to 90% (Figure 3a), reproducible across multiple devices ($n = 3$) (Figure 3b).

Figure 3. Experimental duty cycles overlap predicted values; flow rate ratio manipulation stably and reproducibly regulates duty cycle across multiple devices. **(a)** Filled symbols represent duty cycle values observed and averaged across four syringe combinations and at five different total volumetric inflow rates. Unfilled blue circles represent predicted duty cycle values. All values are derived from time series data containing >6 oscillations. Duty cycle values are plotted against the squared ratio between syringe diameter (*Syringe 2:Syringe 1*) to illustrate the general trend observed. **(b)** Duty cycle data collected from multiple devices ($n = 3$) is presented against the squared ratio between syringe diameter (*Syringe 2:Syringe 1*). Filled symbols represent duty cycle values recorded and averaged across four syringe combinations for total volumetric inflow rates ranging from 5 to 40 µL/min. Unfilled circles represent theoretical (predicted) duty cycle values. All averaged values are derived from time series data containing >6 oscillations. Error bars represent the calculated standard deviation for all duty cycle values recorded from each of three devices for all tested inflow rate ranges.

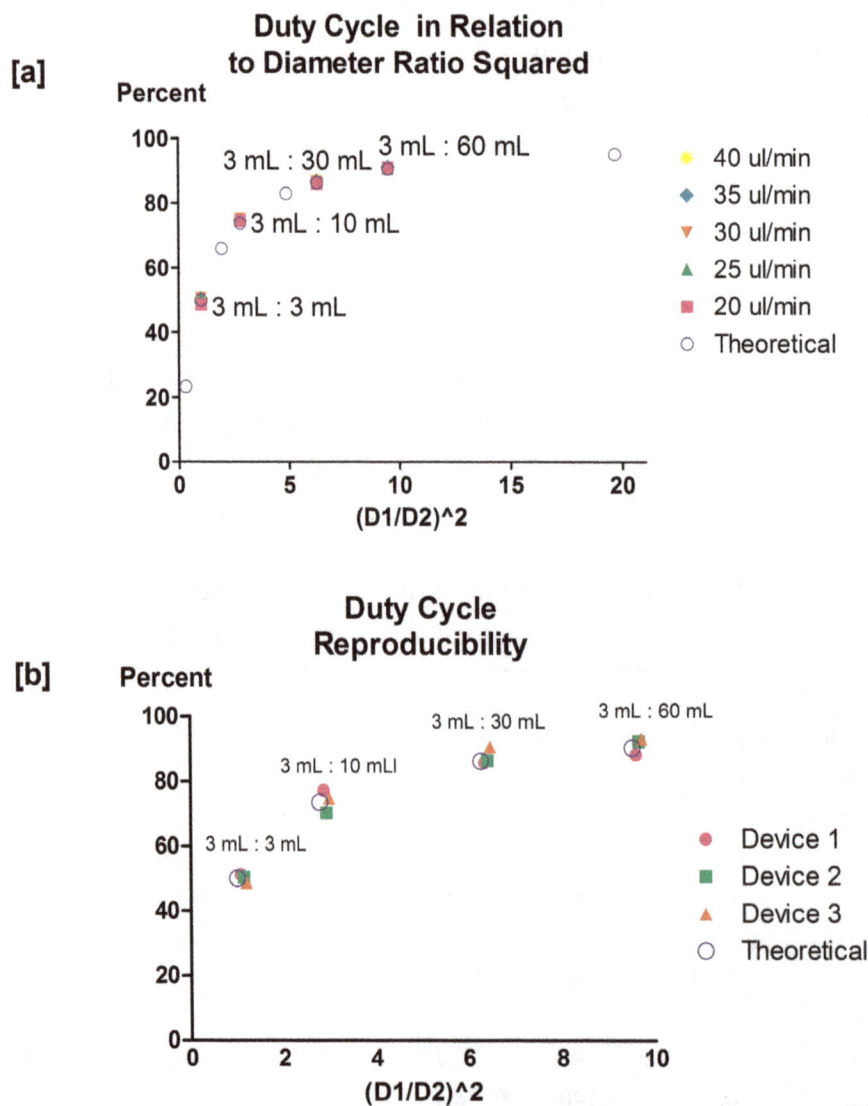

4.2. Mounting Syringes on Separate Syringe Pumps Produces Unstable Duty Cycles

To verify that Equation (4) did not accurately predict duty cycles produced through the sole modification of volumetric flow rate; two identical syringes were mounted on two independent syringe pumps and tested for their ability to produce predictable duty cycles. Flow was initiated at a total volumetric flow rate of 20 µL/min, and the resulting duty cycles were recorded and analyzed. The duty cycles produced via this setup deviated from their predicted values and were unstable, appearing to shift sporadically from one oscillation pattern to another, interspersed by brief periods during which the oscillations would appear stable. This instability was also present at additional total volumetric inflow rates (data not shown), and ultimately affected the predictability of the duty cycles produced (Figure 4).

Figure 4. Single syringe pump setup results in more robust duty cycle control than two pump setup. A minimum of 7 sequential oscillations were observed using two experimental setups (either comprised of a multiple syringes mounted upon a single pump or single syringes mounted upon multiple pumps) to identify reproducibility and consistency of duty cycle. The data presented was acquired using both experimental setups at a total volumetric inflow rate of 20 µL/min. Error bars represent the 95% confidence intervals for experimentally observed results. Two different syringe pump models were utilized in the multiple syringe pump setup.

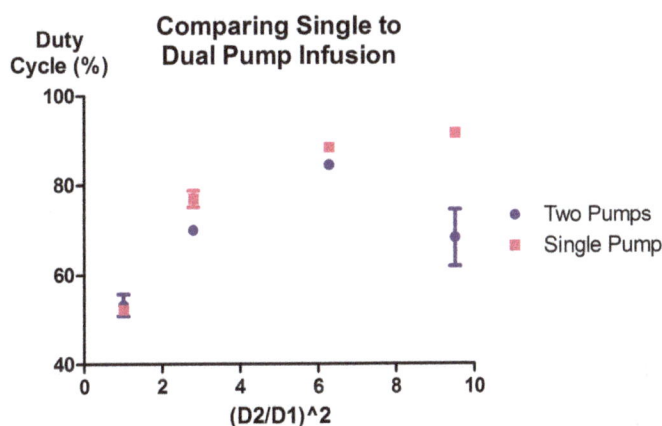

The sources of the observed deviation and instability are likely two-fold. The deviation likely arises as a consequence of the asymmetric linear pressures experienced by each syringe that result in a change in relative P_{th}, but not in C; necessary for performing the reduction yielding Equation (4), and consequently, for the simplified and accurate prediction of duty cycle. The source of the observed instability at a specific flow rate ratio may be multifaceted; deriving from differences in manufacturing of the pumps themselves, differences in their calibration or age, and general unsteadiness inherently observed in syringe pumps [22,26]. As the presence of variability between syringe pumps is unavoidable, the use of multiple syringe pumps presents an inherent risk that predictability of the resulting duty cycle will be adversely effected due to an uncoupling between the pump-derived variability experienced by each individual syringe. Mounting multiple syringes upon a single syringe pump, however, ensures that each syringe experiences similar pump-derived variability. This coupling then ensures that slight instabilities in linear output are experienced simultaneously by both syringes; resulting in a predictable and stable duty cycle.

4.3. Maximum Pressure Profile Remains Relatively Constant

The use of asymmetric inflow rates generated by mounting two syringes of varying diameter onto a single syringe pump alters the pressure profiles generated from each valve (Figure 2b). As we are unable to directly measure gate pressure within our experimental system, we use the previously established approximation, where $P_{S2} \cong P_{G1}$ and $P_{S1} \cong P_{G2}$ at the time of an open-to-close transition [2]. By this approximation, we conclude the asymmetric P_{th} values observed, even under extreme asymmetric conditions ($|P_{th1} - P_{th2}| < 2$ kPa), are far below those reported in previous work ($|P_{th1} - P_{th2}| < 55$ kPa) utilizing asymmetric valve units [8].

The P_{max} values recorded for each valve under the examined flow conditions are equivalent under symmetric volumetric inflow rates, but diverge from these values as the asymmetry between the two inflow rates is increased (Table 2).

Because the transition of each valve from a closed-to-open state is triggered by the accumulation of sufficient fluidic pressure (P_{max}); the initial outflow velocity from each valve is higher (Q_{max}) relative to the stabilized baseline velocity subsequently achieved [24]. The lower P_{max} values observed within this system, relative to values previously-reported [2], suggests a reduction in Q_{max} and, thus, in the magnitude of the transient fluctuation in flow velocity accompanying the transition of each valve from a closed-to-open state. Despite this reduction, as fluidic shear is known to influence the morphological and phenotypical properties of cultured cells and tissues, the mere presence of this fluctuation may nonetheless represent a parameter which must be considered when utilizing this device for the performance of biological analyses.

A comparison of P_{max} values across both valves in one device demonstrates P_{max} values for *valve 1* increase relative to P_{max} values for *valve 2* in proportion to the degree of asymmetry between the inflow rate ratios across the two valves. All data presented is derived from one device, as inter-device variability led to differing absolute P_{max} values across devices. Similar trends, however, were observed across all devices examined.

Table 2. Larger maximum pressures observed in valve receiving smaller inflow rate.

Total Volumetric Inflow Rate (µL/min)	3 mL:3 mL		3 mL:10 mL		3 mL:30 mL		3 mL:60 mL	
	Valve 1 (kPa)	Valve 2 (kPa)	Valve 1 (kPa)	Valve 2 (kPa)	Valve 1 (kPa)	Valve 2 (kPa)	Valve 1 (kPa)	Valve 2 (kPa)
20	3.30	3.34	3.49	3.32	3.51	2.76	3.67	3.04
25	4.08	4.16	4.20	4.01	4.28	3.48	4.53	3.80
30	4.77	4.92	5.03	4.83	5.20	4.28	5.44	4.58
35	5.52	5.70	5.72	5.54	6.11	5.08	6.34	5.35
40	6.31	6.52	6.52	6.29	6.98	5.86	7.18	5.94

4.4. Syringe Properties Influence Capacitance

Syringe size has previously been shown to impact overall compliance in a syringe-driven system, where, independent of material and design, increases in syringe diameter are correlated with increases in syringe compliance [27]. This effect, underappreciated within the field of microfluidics, was observed within our experimental system (Figure 5), and presented a source for concern, as external capacitance could influence the period of the oscillatory output [24]. The good agreement between the duty cycles predicted by the simplified Equation (4) and the actual observed duty cycles are explained by looking at Equation (3), where there is an approximate inverse relationship between C and P_{th} observed under asymmetric inflow rates (described in greater detail below).

Figure 5. Fluidic capacitance increases significantly with increasing syringe volume. Capacitance values were averaged for individual syringes using data collected at multiple volumetric flow rates (ranging from 10 to 40 µL/min). All values are derived from time series data containing >6 oscillations, with five replicates ($p < 0.0002$). Error bars represent the 95% confidence intervals of all capacitance values obtain over multiple inflow rate ranges.

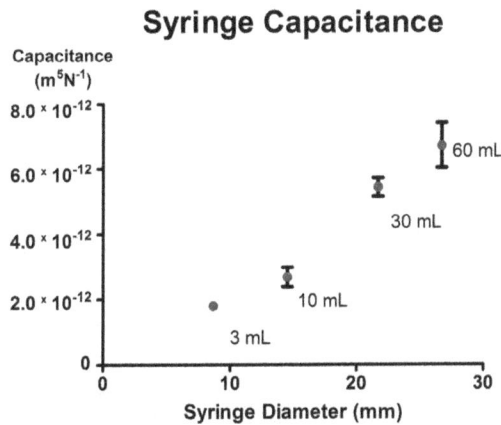

4.5. Different Asymmetric Inflow Rates at Constant Total Volumetric Inflow Rate Produce Distinct Periods

Previous work provides an approximation of the off-time for each valve that can be used to estimate oscillatory period [2], thereby assisting in contextualizing any observed shift in period:

$$t_{off-i} = \left(\frac{C}{Q_{in}}\right) \times P_{th-i} \tag{5}$$

We calculated P_{th} using experimental data collected under multiple inflow conditions. We found that under asymmetric flow regimes, P_{th} and C exhibit an inverse relationship, where P_{th} is higher for the valve experiencing the lower flow rate (*valve 1*), lower for the valve experiencing the higher flow rate (*valve 2*) and where the absolute difference between P_{th} (i.e., $|P_{th1} - P_{th2}|$) increases with the degree of asymmetry between the syringes used. As C is proportional to the size of a given syringe, it is consequently proportional to Q_{in}, which increases with the size of the syringe used. This finding is in agreement with previous results reported for four-way valves, where an increase in volumetric inflow rate through one valve increases calculated P_{th} for the opposite valve [21].

From Equation (5), we infer that increasing P_{th} in conditions with lower Q_{in}, will produce higher t_{off}; and that as the asymmetry between the flow rate across each valve increases, t_{off} will increase for the valve with a lower inflow rate, producing larger oscillation periods.

Using the averaged values of P_{th} and C for each respective syringe pairing, we approximated t_{off} for both valves. We then compared the calculated period approximation with experimental data (Figure 6), and observed that the relationship between volumetric flow rate and period is preserved. We limited the presented period data to one device, as all devices tested exhibited similar trends, with slight variations in absolute values. Such variations may originate from differences in device size (e.g., thickness of the

PDMS membrane), fabrication procedure or material batch characteristics. In addition, larger standard deviations in the period, prominent at greater asymmetric inflow rates, may also originate from fluctuations in syringe pump pressure [26].

This observation highlights the utility of our approach and underscores the motivation for this work. Mounting two syringes of the same size on two independent syringe pumps and evacuating them at two different volumetric flow rates will produce changes in P_{th}, but not in C, introducing a source of complexity to the relationship between volumetric flow rate ratio and duty cycle. Practically, this would result in the inability to reduce down to Equation (4). However, by utilizing syringes of differing diameter, volumetric flow rate-dependent changes in P_{th} are counteracted, allowing one to perform straightforward prediction of duty cycle as a function of volumetric flow rate ratio.

Figure 6. Asymmetric inflow rates produce markedly different periodicity, yet can be estimated relatively-well. Observed period values for each syringe combination demonstrate the range of periodicities generated for each of the four combinations tested. The estimated oscillatory period was calculated by applying Equation (5) for each syringe combination; values for C and P_{th} were derived from the minimal and maximal volumetric inflow rates tested, and were used to establish a linear relationship for P_{th-i} where $P_{th-i} = m \times Q_i + b$. Predicted period values (unfilled) were then compared to the averaged measured period values (filled). All values are derived from time series data containing >12 oscillations, and error bars represent 95% confidence intervals for experimentally observed results.

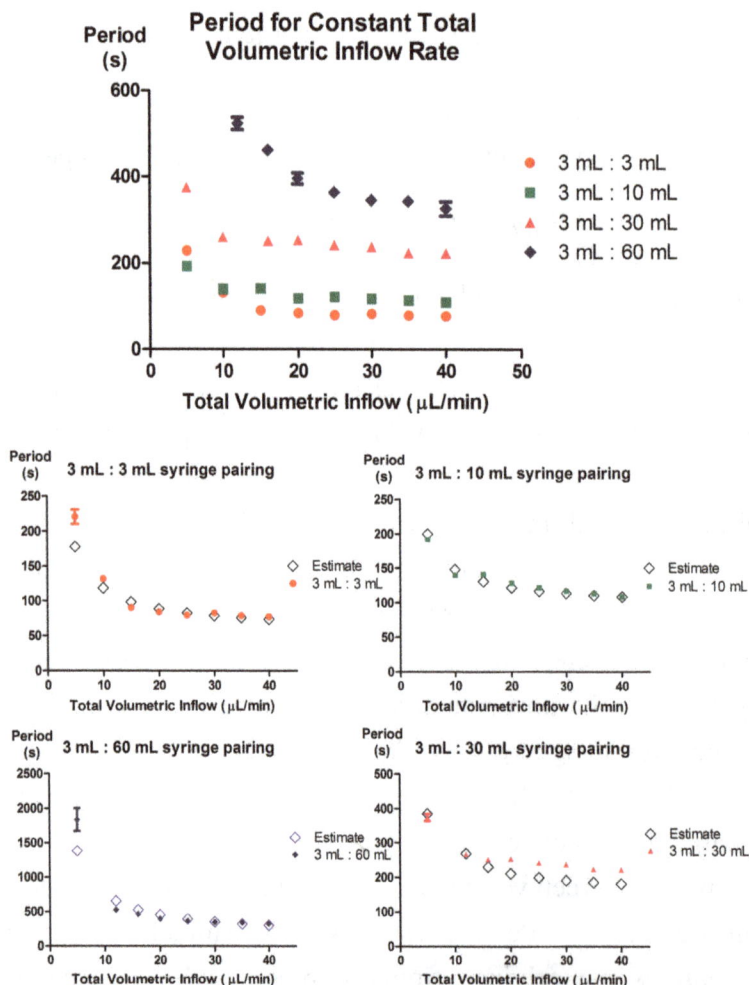

4.6. Estimating Rest and Stimulation Pulse Duration for Control of Rhythmic Stimulation

The described microfluidic oscillator is designed to translate two independent fluid inputs into a single oscillatory fluid output. In practice, if one input contains a fluid stimulant, and the other a neutral "wash" solution, this system may be utilized to conduct biological experiments in which a population of cells (or tissue explant) cultured downstream is presented with this fluid stimulant at a fixed concentration, and for a pre-determined period of time—referred to as the stimulation duration (D); followed by a "wash"—or rest period (R). The functional significance of the presented asymmetric operating technique is that it allows the user to dynamically control the duty cycle of this oscillatory outflow, and in doing so, to characterize biological responses to multiple stimulation regimes characterized by variations in D and R (e.g., fixed D separated by variable R). Within a biological context, control of these parameters is critical as both have been reported to elicit distinct cellular responses [19].

Within the context of the presented device system, D and R may be calculated as a function of relative inflow rates. To do so, C and P_{th} for each valve must be measured with respect to its corresponding syringe and input Q_i values, respectively. Measurements of P_{th} for each valve must be conducted at two total volumetric inflow rates (we used 5 μL/min and 40 μL/min, the minimal and maximal total volumetric inflow rates, respectively) to approximate the linear relationship $P_{th-i} = m \times Q_i + b$. This relationship may then be used to approximate intermediate P_{th-i} values for different inflow rates, and for each syringe pairing. The P_{th-i}, C, and Q_i values may then be used, in equation (5), to determine the off-time for each valve. The sum of the off-times will estimate the periodicity of the device for a given syringe combination. By this method, a curve in general agreement with empirical data, and representing the periodicity as a function of the ratio between syringe diameters, may be generated (Figure 6). This curve may then be utilized to identify an appropriate total volumetric inflow to produce a desired D and R for the specific syringe combination being used. Conversely, this curve may also be utilized to identify the appropriate combination of syringes necessary to modify the length of D or R.

5. Conclusions

The volumetric flow-regulated microfluidic oscillator system described herein greatly increases the versatility and utility of our previously described micro-machine as a tool for generating and delivering pulsatile stimulation. Furthermore, in allowing users to reliably produce a desired duty cycle through the simple manipulation of volumetric inflow rate, the system described greatly reduces the barrier for adoption otherwise presented by placing the burden for "programming" the device upon the end-user. Notably, the benefit of using one syringe pump to drive both syringes is that inherent syringe pump unsteadiness and subsequent inflow fluctuations are applied to both syringes simultaneously; negating their impact on duty cycle, and resulting in a more consistent and stable oscillation.

Acknowledgments

The authors would like to thank the National Institutes of Health (GM096040), the National Science Foundation Graduate Research Fellowship Program and the National Institutes of Health Cellular Biotechnology Training Program (to Sasha Cai Lesher-Perez) under Grant No. DGE 1256260 (ID: 2011101670) and Grant No. NIH GM008353, respectively; as well as the Wayne State University

Interdisciplinary Biomedical Sciences Competitive Research Fellowship (to Priyan Weerappuli); and the National Natural Science Foundation of China (No. 51136007), the Natural Science Foundation of Chongqing, China (No. cstc2013jjB9004), the Research Project of the Chinese Ministry of Education (No. 113053A) and the China Scholarship Council (to Chao Zhang); for financial support.

Author Contributions

Sasha Cai Lesher-Perez, Priyan Weerappuli, Sung-Jin Kim and Shuichi Takayama conceived of and designed the experiments. Sasha Cai Lesher-Perez and Priyan Weerappuli performed all experiments. Sasha Cai Lesher-Perez, Priyan Weerappuli, Sung-Jin Kim and Chao Zhang analyzed and interpreted the data and results. Sasha Cai Lesher-Perez, Priyan Weerappuli and Shuichi Takayama wrote the paper. Sasha Cai Lesher-Perez, Priyan Weerappuli, Sung-Jin Kim, Chao Zhang and Shuichi Takayama assisted in the preparation and approval of the submitted manuscript.

Conflicts of Interest

The authors declare no conflict of interest.

References

1. Mosadegh, B.; Kuo, C.H.; Tung, Y.C.; Torisawa, Y.S.; Bersano-Begey, T.; Tavana, H.; Takayama, S. Integrated elastomeric components for autonomous regulation of sequential and oscillatory flow switching in microfluidic devices. *Nat. Phys.* **2010**, *6*, 433–437.
2. Kim, S.J.; Yokokawa, R.; Lesher-Perez, S.C.; Takayama, S. Constant flow-driven microfluidic oscillator for different duty cycles. *Anal. Chem.* **2012**, *84*, 1152–1156.
3. Walker, G.M.; Beebe, D.J. A passive pumping method for microfluidic devices. *Lab Chip* **2002**, *2*, 131–134.
4. Jamieson, J.D.; Palade, G.E. Intracellular transport of secretory proteins in the pancreatic exocrine cell. I. Role of the peripheral elements of the Golgi complex. *J. Cell Biol.* **1967**, *34*, 577–596.
5. Jamieson, J.D.; Palade, G.E. Intracellular transport of secretory proteins in the pancreatic exocrine cell. II. Transport to condensing vacuoles and zymogen granules. *J. Cell Biol.* **1967**, *34*, 597–615.
6. Gratzner, H.G. Monoclonal antibody to 5-bromo- and 5-iododeoxyuridine: A new reagent for detection of DNA replication. *Science* **1982**, *218*, 474–475.
7. Dolmetsch, R.E.; Xu, K.; Lewis, R.S. Calcium oscillations increase the efficiency and specificity of gene expression. *Nature* **1998**, *392*, 933–936.
8. Taylor, R.J.; Falconnet, D.; Niemistö, A.; Ramsey, S.A.; Prinz, S.; Shmulevich, I.; Galitski, T.; Hansen, C.L. Dynamic analysis of MAPK signaling using a high-throughput microfluidic single-cell imaging platform. *Proc. Natl. Acad. Sci. USA* **2009**, *106*, 3758–3763.
9. Tay, S.; Hughey, J.J.; Lee, T.K.; Lipniacki, T.; Quake, S.R.; Covert, M.W. Single-cell NF-κB dynamics reveal digital activation and analogue information processing. *Nature* **2010**, *466*, 267–271.
10. Cheong, R.; Wang, C.J.; Levchenko, A. High content cell screening in a microfluidic device. *Mol. Cell. Proteomics* **2009**, *8*, 433–442.

11. Sackmann, E.K.; Fulton, A.L.; Beebe, D.J. The present and future role of microfluidics in biomedical research. *Nature* **2014**, *507*, 181–189.

12. Mosadegh, B.; Bersano-Begey, T.; Park, J.Y.; Burns, M.A.; Takayama, S. Next-generation integrated microfluidic circuits. *Lab Chip* **2011**, *11*, 2813–2818.

13. Leslie, D.C.; Easley, C.J.; Seker, E.; Karlinsey, J.M.; Utz, M.; Begley, M.R.; Landers, J.P. Frequency-specific flow control in microfluidic circuits with passive elastomeric features. *Nat. Phys.* **2009**, *5*, 231–235.

14. Duncan, P.N.; Nguyen, T.V.; Hui, E.E. Pneumatic oscillator circuits for timing and control of integrated microfluidics. *Proc. Natl. Acad. Sci. USA* **2013**, *110*, 18104–18109.

15. Rhee, M.; Burns, M.A. Microfluidic pneumatic logic circuits and digital pneumatic microprocessors for integrated microfluidic systems. *Lab Chip* **2009**, *9*, 3131–3143.

16. Collino, R.R.; Reilly-Shapiro, N.; Foresman, B.; Xu, K.; Utz, M.; Landers, J.P.; Begley, M.R. Flow switching in microfluidic networks using passive features and frequency timing. *Lab Chip* **2013**, *13*, 3668–3674.

17. Toepke, M.W.; Abhyankar, V.V.; Beebe, D.J. Microfluidic logic gates and timers. *Lab Chip* **2007**, *7*, 1449–1453.

18. Kim, S.-J.; Lai, D.; Park, J.Y.; Yokokawa, R.; Takayama, S. Microfluidic automation using elastomeric valves and droplets: Reducing reliance on external controllers. *Small* **2012**, *8*, 2925–2934.

19. Jovic, A., Howell, B.; Cote, M.; Wade, S.M.; Mehta, K.; Miyawaki, A.; Neubig, R.R.; Linderman, J.J.; Takayama, S. Phase locked signals elucidate circuit architecture of an oscillatory pathway. *PLoS Comput. Biol.* **2010**, *6*, e1001040.

20. Jovic, A.; Howell, B.; Takayama, S. Timing is everything: using fluidics to understand the role of temporal dynamics in cellular systems. *Microfluid. Nanofluid.* **2009**, *6*, 717–729.

21. Kim, S.J.; Yokokawa, R.; Takayama, S. Analyzing threshold pressure limitations in microfluidic transistors for self-regulated microfluidic circuits. *Appl. Phys. Lett.* **2012**, *101*, 234107.

22. Korczyk, P.M.; Cybulski, O.; Makulska, S.; Garstecki, P. Effects of unsteadiness of the rates of flow on the dynamics of formation of droplets in microfluidic systems. *Lab Chip* **2011**, *11*, 173–175.

23. Rooke, G.A.; Bowdle, T.A. Syringe pumps for infusion of vasoactive drugs: mechanical idiosyncrasies and recommended operating procedures. *Anesth. Analg.* **1994**, *78*, 150–156.

24. Kim, S.J.; Yokokawa, R.; Takayama, S. Microfluidic oscillators with widely tunable periods. *Lab Chip* **2013**, *13*, 1644–1648.

25. Peakdet: Peak Detection Using MATLAB. Available online: http://billauer.co.il/peakdet.html (accessed on 24 November 2014).

26. Li, Z.; Mak, S.Y.; Sauret, A.; Shum, H.C. Syringe-pump-induced fluctuation in all-aqueous microfluidic system implications for flow rate accuracy. *Lab Chip* **2014**, *14*, 744–749.

27. Weiss, M.; Hug, M.I.; Neff, T.; Fischer, J. Syringe size and flow rate affect drug delivery from syringe pumps. *Can. J. Anaesth.* **2000**, *47*, 1031–1035.

9

Fast Prototyping of Sensorized Cell Culture Chips and Microfluidic Systems with Ultrashort Laser Pulses

Sebastian M. Bonk [1], Paul Oldorf [2], Rigo Peters [2], Werner Baumann [1] and Jan Gimsa [1,*]

[1] Chair of Biophysics, University of Rostock, Rostock 18057, Germany;
    E-Mails: sebastian.bonk@uni-rostock.de (S.M.B.); werner.baumann@uni-rostock.de (W.B.)

[2] SLV Mecklenburg-Vorpommern GmbH, Rostock 18069, Germany;
    E-Mails: oldorf@slv-rostock.de (P.O.); peters@slv-rostock.de (R.P.)

* Author to whom correspondence should be addressed; E-Mail: jan.gimsa@uni-rostock.de

Academic Editor: Nam-Trung Nguyen

Abstract: We developed a confined microfluidic cell culture system with a bottom plate made of a microscopic slide with planar platinum sensors for the measurement of acidification, oxygen consumption, and cell adhesion. The slides were commercial slides with indium tin oxide (ITO) plating or were prepared from platinum sputtering (100 nm) onto a 10-nm titanium adhesion layer. Direct processing of the sensor structures (approximately three minutes per chip) by an ultrashort pulse laser facilitated the production of the prototypes. pH-sensitive areas were produced by the sputtering of 60-nm Si_3N_4 through a simple mask made from a circuit board material. The system body and polydimethylsiloxane (PDMS) molding forms for the microfluidic structures were manufactured by micromilling using a printed circuit board (PCB) milling machine for circuit boards. The microfluidic structure was finally imprinted in PDMS. Our approach avoided the use of photolithographic techniques and enabled fast and cost-efficient prototyping of the systems. Alternatively, the direct production of metallic, ceramic or polymeric molding tools was tested. The use of ultrashort pulse lasers improved the precision of the structures and avoided any contact of the final structures with toxic chemicals and possible adverse effects for the cell culture in lab-on-a-chip systems.

Keywords: rapid prototyping; micro sensor chip; ITO; oxygen; pH; picosecond laser; cell monitoring system; top-down approach

1. Introduction

The development of multifunctional microfluidic systems calls for the integration of sensors and actuators such as electrical sensors and heating or pump elements, all of which are routinely manufactured using well-established microelectronic fabrication techniques. Microfluidic components are also produced using photolithographic technology, especially in basic research [1]. Nevertheless, injection molding is unrivalled amongst the various polymer technologies used in the mass production of microfluidic components. Microinjection molding, hot embossing, or hot shaping are alternatives for low volume production or prototyping [2].

The materials used in biotechnological and medical applications must comply with certain requirements, e.g., biocompatibility or stability against heat sterilization. These requirements are challenging with respect to the physical and chemical properties, especially in lab-on-a-chip devices for long-term cell monitoring or for micro-reaction techniques in Micro Total Analysis Systems (μTAS) [3]. Unfortunately, glass, as the material of choice, requires special adaptations in the production methods [4]. For low volume production or prototyping, photolithography techniques are time-consuming and costly due to their high number of processing steps. Often, the outdated (4-inch) disk-wafer technology is used, which restricts the size and the degrees of freedom in the individual chip design [5].

To increase the frequency of prototype realization and for small batch series (a requirement in basic research), it is necessary to implement fast design cycles using flexible and fast production technologies. This strategy shortens development times and test cycles in the research-intensive field of microfluidic systems development. As a consequence, rapid prototyping techniques are gaining increasing importance because they help reduce costs through more efficient employment of human resources.

To our knowledge, lasers are routinely used in the production of precise circuit board structures as small as 50-μm wide [6,7], whereas the photolithographic technology dominates thin layer structuring, especially of platinum or ITO layers on silicon or glass [8,9]. With respect to the conventional long-pulse lasers, ultrashort pulse lasers improve the quality of ablation because a higher percentage of their pulse energy is employed in evaporation instead of melting processes [10,11]. This is another reason why the technology is gaining importance in the surface modification of thin layers [12] or in the industrial production of solar cells [13].

In some cases, laser techniques were used by other authors for the production of channels or recesses in microfluidic systems [4,14–16].

Here, we present a new prototyping technology for sensorized microfluidic cell-culture systems, which is based on a combination of different applications of an ultrashort pulse laser system. The system consists of a polycarbonate body with inlets and outlets for fluids, as well as a polycarbonate lid (*cf.* Section 2.4). The lid bears the microfluidic structure, which was imprinted in polydimethylsiloxane (PDMS). In the assembled system, the microfluidic structure is sealed by a microscopic glass slide, which bears the thin layer-sensor structures.

The polycarbonate body was produced using a standard printed circuit board (PCB) micromilling machine. Here, this machine was also used to produce the PDMS molding tools for imprinting the microfluidic structures into the PDMS lid. More precise, high temperature-stable molding tools, which are needed for hot embossing, injection molding or hot shaping, could be produced from ceramic wrought materials (Keralpor99, Kerafol GmbH, Eschenbach, Germany). Direct laser structuring is an alternative to the common photolithographic production of PDMS molding tools, e.g., from SU-8 or silicon [1,17]. In our system, the microfluidic structure was confined by a microscopic glass slide on the bottom. The slide carried amperometric Clark oxygen sensors, potentiometric pH and impedimetric cell adhesion sensors interdigitated electrode structures (IDES) that were originally developed with photolithographic techniques and redesigned for laser structuring after testing [18].

Our new technology circumvents the use of clean rooms and toxic or harmful chemicals, thereby shortening manufacturing times. This technology allows for the use of a wide variety of wafer formats. Here, we present sensorized chips that were produced from cheap microscopic glass slides. Together with a microfluidic PDMS structure, they allow for long-term (days to weeks) cell culture and on-line registration of sensor readouts.

2. Experimental Section

2.1. Laser Structuring of Platinum and ITO Layers

Microscope slides (NK72.1, Carl Roth, Karlsruhe, Germany) were cleaned with aqueous 10% Tickopur W77 cleaning solution (BANDELIN electronic GmbH & Co. KG, Berlin, Germany) in an ultrasonic bath and dried in a pressurised air stream. The slides were sputtered with 130 nm of platinum on a 15-nm titanium undercoating using a magnetron sputter system (Ardenne LA-320S, VON ARDENNE GmbH, Dresden, Germany). The chips were structured by ablation with a picosecond laser (TruMicro 5X50, TRUMPF Laser- und Systemtechnik GmbH, Ditzingen, Germany) using a high-precision micromachining system (GL.5, GFH GmbH, Deggendorf, Germany). The laser could be switched between three wavelengths, 1030 nm (fundamental frequency), 515 nm (second harmonic frequency), and 343 nm (third harmonic frequency). The positioning precision of the system was better than 1 μm. The structure designs were drawn in AutoCAD 2010 (Autdesk, Inc., San Rafael, CA, USA) and directly exported into the laser control system as DXF-files [19]. For chip structuring, platinum or ITO (CEC020, Präzisions Glas & Optik GmbH, Iserlohn, Germany) were ablated from the microscope slide at a wavelength of 343 nm with a pulse frequency of 400 kHz with a pulse energy of 1.25 μJ, and an effective spot diameter of 10 μm. At 1000 mm/s driving, the pulse spot overlapped approximately 80%. To ensure the microscopic transparency of the chip areas without sensor structures, platinum areas were ablated in bidirectional stripes of 8-μm distance. This ablation step could be omitted in ITO structuring (Figure 1). For both materials, the sensor structures were outlined in a final step at 100 mm/s driving, a pulse frequency of 40 kHz and a pulse energy of 1.25 μJ. To compensate for the size of the laser spot, the outlines of the structures were enlarged by 6 μm in the AutoCAD files.

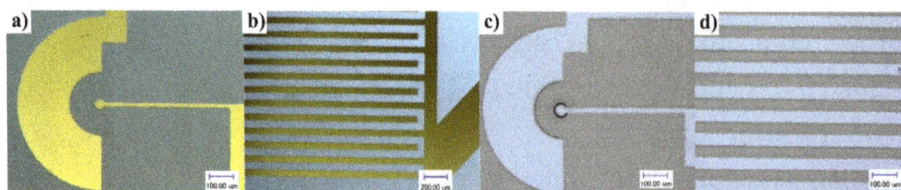

Figure 1. Platinum (**a,b**) and indium tin oxide (ITO) (**c,d**) structures of the oxygen sensors with a circular working electrode spot and semi-circular counter electrodes (**a,c**), as well as interdigitated electrode structures (IDES) with a50-μm pitch (**b,d**). The on-chip-sensor connectors were insulated in a following processing step.

2.2. Laser Structuring of Silicon Nitride (Si₃N₄)

After structuring, the chips were cleaned again, as described above, and sputtered with a 200-nm Si_3N_4 passivation layer (300 s at 400 W in high frequency mode). For a thicker passivation layer, another chip batch was passivated with 1000 nm of Si_3N_4 (GeSiM mbH, Grosserkmannsdorf, Germany) using a PE-CVD process. Si_3N_4 was chosen because of its chemical, electrical and optical characteristics, as well as its biocompatibility. A second laser-processing step was executed to create windows in the passivation layer for the external chip connectors and the sensors. Si_3N_4 layers of both thicknesses were ablated at 1000 mm/s driving, a pulse frequency of 200 kHz and a pulse spot overlap of approximately 70%. To remove the Si_3N_4 passivation layer only and not the platinum layers, a maximum pulse energy of 1.0 µJ was used. To produce pH sensors in a final step, a 60-nm Si_3N_4 layer was sputtered through four openings in a PCB, which was used as a cheap and simple mask.

2.3. Laser Structuring of Molding Tools by Laser Ablation

In an attempt to overcome the limitations of the stereo-lithographic (material, aspect ratio) and micro-milling techniques (precision, surface roughness), the laser system was used to create new molding tools for microfluidics. A wrought aluminum oxide ceramic chip was laser processed in the same system with at a wavelength of 1030 nm, a pulse energy of 125 µJ, 2000 mm/s driving and a pulse rate of 200 kHz. A structure depth of 240 µm was reached after 20 repetitions.

2.4. Cell Culture System and Microfluidics Fabrication

Figure 2 presents the cell culture system consisting of a body with fluidic connectors and a lid with a central PDMS membrane. An outside ring and inner spacer structures confined the microfluidic cell culture volume. All of the polycarbonate parts, i.e., the system's body, lid and the PDMS molding tools for the microfluidic structures, were produced by a PCB micromilling machine (CCD 2, BUNGARD Elektronik, Windeck, Germany). The body and lid were designed in AutoCad and milled in an 8-mm polycarbonate plate, whereas the molding tools were milled in polymethyl methacrylate (PMMA) because of its better grindability with smaller milling tools (<500 µm in diameter). For the larger fluidic body structures of the cell culture system, Teflon® could also be used for the molding tools. Because of its low surface adhesion, Teflon® has advantages in the separation of the PDMS parts from the casting molds, though it is delicate to mill with small tools.

To produce the PDMS parts, PDMS monomer (Sylgard 184, Dow Corning Inc. Midland, MI, USA) was degassed, injected into the molding tools and cured for 4 h at 70 °C. After removing the molded parts, the bottom side of the PDMS body was manually coated with a very thin layer of uncured PDMS before the body was fixed to the chip under pressure and cured for a second time. For improved PDMS adhesion, the polycarbonate parts and the chip surfaces were coated with a bonding agent (GRUN G790, Wacker Chemie AG, München, Germany) before PDMS injection. After curing the PDMS, the microfluidic connectors were glued to the body. The connectors were shortened syringe needles fixed with their plastic connectors in cross-shaped notches of the body (Figure 2). The notches hindered rotation of the syringe heads and facilitated the attachment of Luer lock connectors. The removable lid enabled access to the culturing volume and the integration of 3D scaffolds.

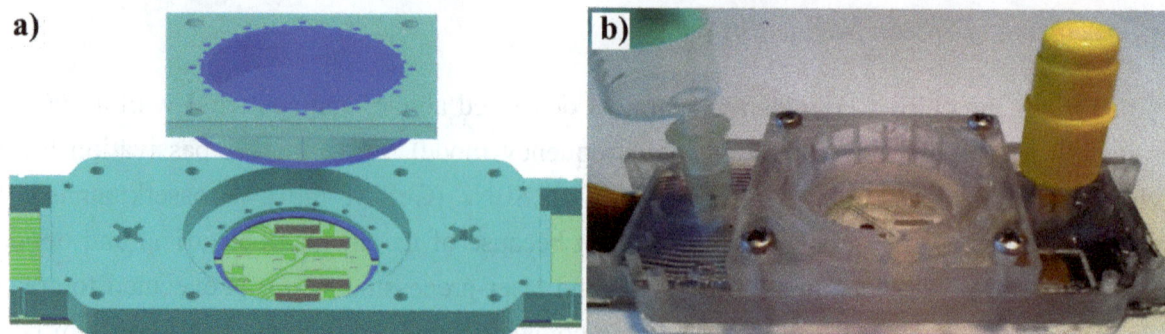

Figure 2. Computer-aided design of the microfluidic system (**a**) with the body (bottom) and lid (top) milled in polycarbonate (light turquoise). The blue polydimethylsiloxane (PDMS)-imprinted parts (gasket with fluidic channels in the body and PDMS membrane in the lid) confine the fluidic volume. The laser-structured chip (green) is glued to the lower PDMS surface. (**b**) The assembled cell culture system with a flexible printed circuit board connector (lower left), as well as a syringe and syringe-injection cap on the left and right inlets, respectively.

2.5. Cell Culture

To verify the biocompatibility, a mouse osteoblast precursor cell line (MC3T3-E1) was used. These cells, which are a common model system for bone regeneration, were obtained from the German collection of microorganisms and cell culture (DMSZ GmbH, Braunschweig, Germany). The cells were cultured in an incubator at 37 °C with 95% humidity and 5% CO_2. The cells were grown to confluence in 50-mL cell culture flasks (25 cm^2; Greiner bio-one, Frickenhausen, Germany) in alpha medium (order No. F 0925) supplemented with 1% penicillin/streptomycin (stem solution: 100 U/mL penicillin and 100 μg/mL streptomycin) and 10% foetal bovine serum (all purchased from Biochrom AG, Berlin, Germany). After the cells grew to confluence, they were trypsinated in phosphate-buffered saline supplemented with 0.05% trypsin and 0.02% EDTA (PAN Biotech GmbH, Aidenbach, Germany) and were diluted before subculture or transfer to the microfluidic systems.

For cell culture in the microfluidic systems, carbonate-free alpha medium (P03-2510, Pan Biotech, Aidenbach, Germany) was supplemented as described above and buffered with 20 mM HEPES (Carl Roth, Karlsruhe, Germany). For cell seeding, 500 μL of the cell suspension with up to 500,000 cells/mL (corresponding to 50,000 cells per culture system) were injected by slow flushing of the system. Then, the system was closed with a medium-filled 6-mL syringe at the inlet and a syringe injection cap as a cannula injection port at the outlet (Figure 2). For medium exchange, an empty, open syringe with a 0.6-mm cannula was plugged into the injection port, and 500 μL of medium from the filled syringe was slowly flushed through the system.

3. Results and Discussion

3.1. Structuring of Platinum and ITO

For processing, the sputtered chips were aligned with the limit stop of the laser workbench. The determination of the optimal laser parameters (power, driving, pulse rate, and pulse spot overlap) for structuring platinum or ITO layers required several pre-runs. Less than 3 min per chip were needed with the optimized laser process (chip area of 11.5 cm^2). The structures could be manufactured with widths less than 20 μm and minimal distances of 10 μm. The smallest on-chip structures were the working electrodes of the oxygen sensors with 40-μm diameter and a conducting path with 20-μm width.

The edge roughness of the structures was usually in the range of ±1.5 μm. We found that this result was less due to a problem of the beam guidance but primarily a consequence of the inhomogeneous ablation in the outer zones of the laser spot where the ablation depended on the trace curvature and the materials (Figure 3).

As observed in Figure 4, the distances between the IDES fingers were approximately 2 μm wider than expected, whereas their width was approximately 2 μm too small. This small mismatch between the design and manufactured structure resulted from an insufficient offset correction.

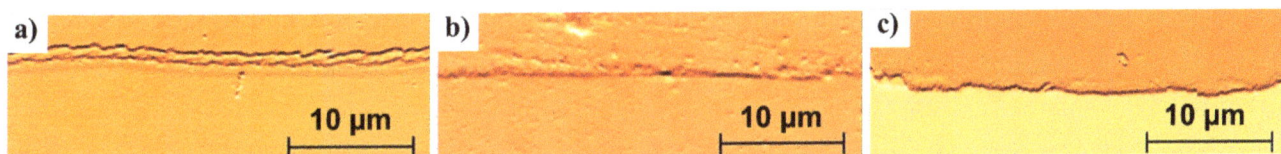

Figure 3. Edge quality of an IDES finger controlled by colour contrasted atomic force microscopic images. (**a**) ITO on glass; (**b**) platinum on glass; (**c**) photolithographically (photo mask) produced chip (GeSiM®).

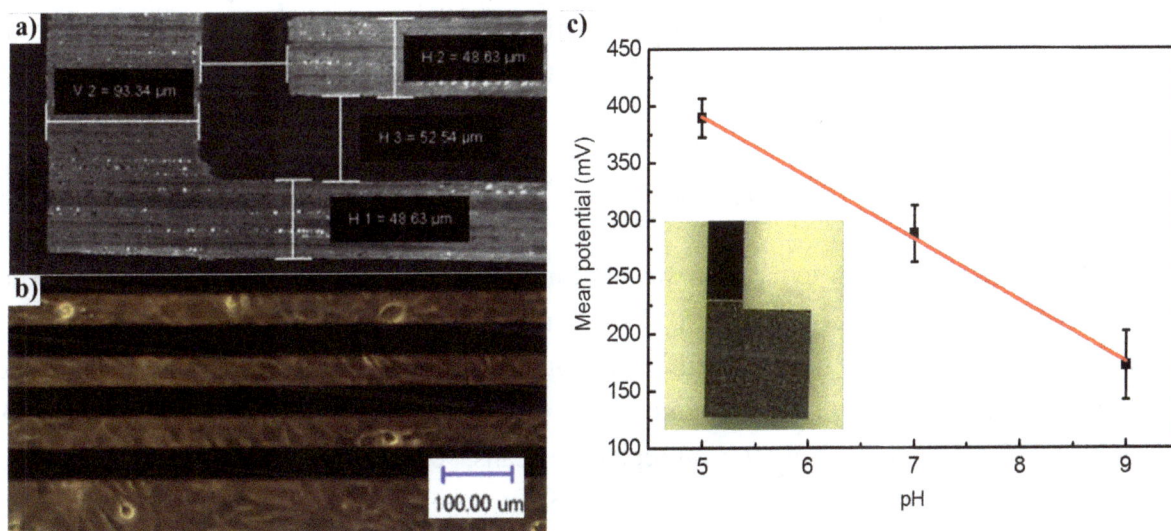

Figure 4. Images of a platinum IDES. (**a**) Scanning electron microscopy image with dimensions. (**b**) Confluent cells above the IDES. (**c**) Sensitivity test of a Si$_3$N$_4$ sensor layer on platinum (insert: pH sensor with 1.4 × 1.4 mm^2 electrode area). The data represent measurements with six chips in pH-buffered solutions with pH values of 5, 7, and 9. The straight line corresponds to a sensitivity of −53.8 ± 1.8 mV/pH.

Despite the ease of small offset corrections, small circular structures are more difficult to produce. The effective laser driving, *i.e.*, the linear distance between three pulse spots, is decreased in curved areas, which led to an increased pulse spot overlap, a stronger removal of material, and even some ablation of the glass carrier (see Figure 1c, oxygen working electrode). This problem will either be solved in a preliminary approach by a modification of the laser parameters in the individual software files for curved structures or with the next update of the driver software of the manufacturer of the laser positioning system (GFH GmbH).

3.2. Structuring of Si₃N₄

The laser parameters for processing the Si_3N_4 passivation layer were confined to a very small window by the underlying platinum and ITO structures of the IDES and oxygen sensors or the peripheral chip contacts. Within this window, 200-nm layers of Si_3N_4 could be removed without removing the platinum layer [19]. The ablation efficiency depended on the underlying substrate. The ablation of Si_3N_4 from glass required higher pulse energy than that over platinum because the ablation thresholds for Si_3N_4 were reduced over glass. Nevertheless, for platinum, the process led to roughening of the surfaces due to the punctual ablation (see the IDES in Figure 4). Interestingly, the 200- and 1000-nm Si_3N_4 layers could be removed using the same laser parameters, sustaining the theory of the effects at the Si_3N_4-platinum or Si_3N_4-glass interfaces that influenced the ablation process [13].

To produce pH sensors, 60 nm of Si_3N_4 were deposited on the laser-manufactured platinum structures. The chips were tested for their stabilities and pH sensitivities by measurements in standard buffer solutions with pH values of 5, 7 and 9 against an Ag/AgCl microelectrode (Microelectrode Inc., Bedford, NH, USA). The chips provided reproducible potentials 10 s after medium exchange. They were kept in the solution at room temperature for up to 600 s to confirm the stability of their readouts. The potentials were linearly dependent on the pH in the range from 5 to 9 at a sensitivity of -53.8 ± 1.8 mV/pH. This sensitivity was comparable to the sensitivity of ISFETS with a Si_3N_4-gate passivation layer [20].

3.3. Structuring Molding Tools by Laser Ablation

Laser manufacturing of molding tools from wrought aluminum oxide ceramic chips was fast and provided a very good surface quality (Figure 5). Per ablation scan, a depth of approximately 12 μm was obtained with 2000 mm/s driving and a repetition rate of 200 kHz. For our structures, the area-specific processing time of approximately 3600 mm^2/min per scan largely referred to the actual ablated area. A structure depth of 240 μm was reached after 20 repetitive scans, resulting in a 5 min processing time for the surface of the 30×30 mm^2 moulding tool (Figure 5a).

The moulding tool for the large sensor chip with a final depth of 465 μm was processed with modified parameters (Figure 5b). The driving and repetition rate were increased to 4000 mm/s and 400 kHz, respectively. This modification reduced the manufacturing time to approximately 35 min for 42 scans despite the larger surface of the tool.

The small tilts of the vertical structures were caused by optical effects in the ablating process (see the laser scanning microscope images in Figure 5). We believed that the resulting tilts in the fluidic channel walls would be advantageous in finally stripping the mold from the cast. If necessary, the tilts could be reduced by using special trepanning optics. Nevertheless, the smooth removal of the cured PDMS molds from the ceramic tools required the use of a mold release agent.

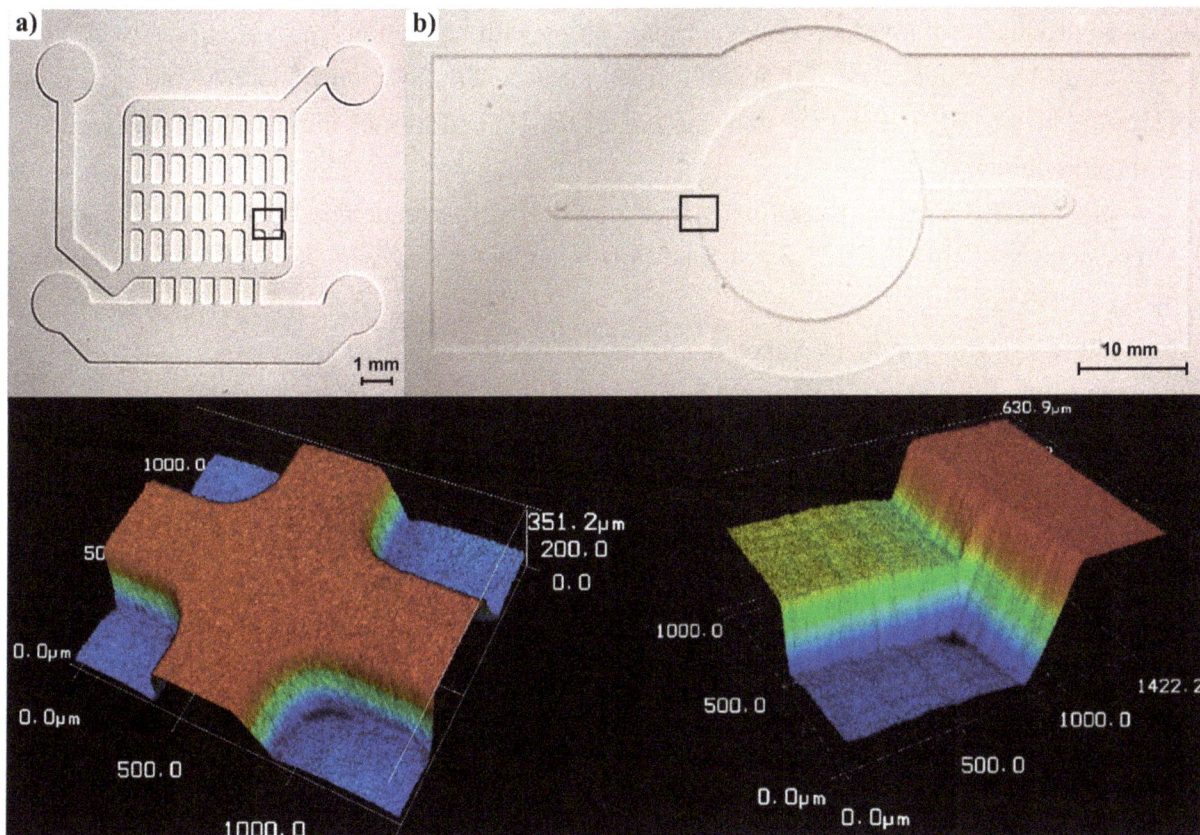

Figure 5. Two ceramic mold chips processed by laser ablation. Black squares mark the positions of the laser scanning microscope images below the mold images. (**a**) A 30 × 30 mm² chip for a microfluidic pattern with 240-μm channel height. (**b**) Larger structure (33 × 76 mm²) for the cell culture system with heights of approximately 230 μm (horizontal inlet and outlet channels) and 465 μm (circular area). Please note that the measures given in the images for the vertical scaling refer to a larger measuring frame.

3.4. Cell culture

MC3T3-E1 cells showed good proliferation in the microfluidic cell culture systems, reaching confluence approximately three days after seeding. The cells adhered to the Si_3N_4, as well as the platinum surfaces without surface coatings, e.g., with poly-D-lysin or laminin [5,18], suggesting that the laser ablation-induced surface roughness may improve cell adhesion [21,22]. Interestingly, the aqueous cell culture medium evaporated through the PDMS lid, resulting in the formation of perturbing gas bubbles. Bubble formation was likely enhanced by the roughness of the cutting edges of the mold structures, which may have been transferred to the PDMS surfaces. In our system, the bubble formation could be suppressed by overlaying the lid of the cell culture system with deionized water, allowing for cell culture times longer than 10 days.

4. Conclusions and Outlook

The use of micromachined bodies and molding tools produced by the well-established PDMS technology in combination with ultrashort pulse laser micromachining of the sensor chips enabled fast prototyping of sensorized microfluidic cell culture systems without any photolithographic processing steps. We could produce thin-layer structures on glass chips with resolutions of approximately 10 μm. The use of this system reduced the fabrication times because the ablation of larger chip areas for microscopy could be spared. The optical transparency of the ITO substrates improved the microscopic analysis of the cell culture.

Fresh pH-sensitive Si_3N_4 layers were finally sputtered on the laser-machined structures. The layers showed a very good sensitivity of -53.8 ± 1.8 mV/pH when compared with ISFETs with Si_3N_4 gate passivation layers [23].

Our cell culture systems could be vapor sterilized and easily modified by remolding the PDMS structures of their lids. MC3T3-E1 cells could be cultured longer than 10 days with good cell adhesion to the laser-manufactured chip surfaces even without further surface modifications. The gas permeability of the PDMS lid ensured 100% air saturation in the culture medium, which may be desirable for the cells but could cause bubble formation. Smoother molds produced, for example, by laser processing, would likely reduce this problem in future microfluidic designs.

Ultrashort pulse laser techniques could also be used to produce casting molds for PDMS molding or even hot embossing of PMMA. This would require robust, temperature-stable casting molds manufactured, for example, from aluminium oxide ceramics. Our prototypes were produced in 3 to 35 min with a resolution of approximately 10 μm in the horizontal direction and approximately 13 μm in the vertical direction. This resolution outperformed that of the currently available stereo-lithographic 3D printers.

For the current system, a machine hour costs approximately €220, resulting in processing costs of €5 per ITO chip in a 60-chip batch. This system also allowed for the production of holes or recesses for advanced microfluidic designs (not shown). The maximum driving of our system of up to 4000 mm/s will be improved in the next generation of ablation systems.

Acknowledgments

The authors are grateful to the German Research Council (DFG) for their funding of the graduate school GRK1505 "WELISA". We would like to thank Dr. Steffen Howitz (GeSiM GmbH, Grosserkmannsdorf) for help in the chip production, as well as Reik Modrozynski (Institute of Biophysics, University of Rostock) for his help with the cell culture. The authors would also like to thank Dipl.-Ing. Jürgen Josupeit and the group of Prof. Dr. Mathias Nowottnick (Institute of Electronic Appliances and Circuits, University of Rostock) for granting access and operating the magnetron sputter system.

Author Contributions

Sebastian M. Bonk designed all experiments with the microfluidic cell culture system. The laser experiments were performed by Paul Oldorf. Jan Gimsa, Werner Baumann, and Rigo Peters contributed knowledge, devices, materials, and reagents. Sebastian M. Bonk provided a manuscript draft. He was financed by grants to Jan Gimsa who finalized the manuscript.

Conflicts of Interest

The authors declare no conflict of interest.

References

1. Chiriacò, M.S.; Primiceri, E.; D'Amone, E.; Ionescu, R.E.; Rinaldi, R.; Maruccio, G. EIS microfluidic chips for flow immunoassay and ultrasensitive cholera toxin detection. *Lab Chip* **2011**, *11*, 658–663.

2. Shah, J.J.; Geist, J.; Locascio, L.E.; Gaitan, M.; Rao, M.V.; Vreeland, W.N. Capillarity induced solvent-actuated bonding of polymeric microfluidic devices. *Anal. Chem.* **2006**, *78*, 3348–3353.

3. Kovarik, M.L.; Gach, P.C.; Ornoff, D.M.; Wang, Y.; Balowski, J.; Farrag, L.; Allbritton, N.L. Micro Total Analysis Systems for Cell Biology and Biochemical Assays. *Anal. Chem.* **2012**, *84*, 516–540.

4. Khan Malek, C.; Robert, L.; Boy, J.-J.; Blind, P. Deep microstructuring in glass for microfluidic applications. *Microsyst. Technol.* **2006**, *13*, 447–453.

5. Koester, P.J.; Buehler, S.M.; Stubbe, M.; Tautorat, C.; Niendorf, M.; Baumann, W.; Gimsa, J. Modular glass chip system measuring the electric activity and adhesion of neuronal cells—Application and drug testing with sodium valproic acid. *Lab Chip* **2010**, *10*, 1579–1586.

6. Zheng, H.; Gan, E.; Lim, G.C. Investigation of laser via formation technology for the manufacturing of high density substrates. *Opt. Lasers Eng.* **2001**, *36*, 355–371.

7. Moorhouse, C.J.; Villarreal, F.; Wendland, J.J.; Baker, H.J.; Hall, D.R.; Hand, D.P. Enhanced peak power CO_2 laser processing of PCB materials. *Proc. SPIE* **2005**, *5827*, 438–444.

8. Ahadian, S.; Ramón-Azcón, J.; Ostrovidov, S.; Camci-Unal, G.; Hosseini, V.; Kaji, H.; Ino, K.; Shiku, H.; Khademhosseini, A.; Matsue, T. Interdigitated array of Pt electrodes for electrical stimulation and engineering of aligned muscle tissue. *Lab Chip* **2012**, *12*, 3491–3503.

9. Yang, L.; Li, Y.; Griffis, C.L.; Johnson, M.G. Interdigitated microelectrode (IME) impedance sensor for the detection of viable Salmonella typhimurium. *Biosens. Bioelectron.* **2004**, *19*, 1139–1147.

10. Sugioka, K.; Cheng, Y. Ultrafast lasers—Reliable tools for advanced materials processing. *Light Sci. Appl.* **2014**, *3*, e149, doi:10.1038/lsa.2014.30.

11. Gattass, R.R.; Mazur, E. Femtosecond laser micromachining in transparent materials. *Nat. Photonics* **2008**, *2*, 219–225.

12. Ahmmed, K.; Grambow, C.; Kietzig, A.-M. Fabrication of Micro/Nano Structures on Metals by Femtosecond Laser Micromachining. *Micromachines* **2014**, *5*, 1219–1253.

13. Bovatsek, J.; Tamhankar, A.; Patel, R.S.; Bulgakova, N.M.; Bonse, J. Thin film removal mechanisms in ns-laser processing of photovoltaic materials. *Thin Solid Films* **2010**, *518*, 2897–2904.

14. Liao, Y.; Song, J.; Li, E.; Luo, Y.; Shen, Y.; Chen, D.; Cheng, Y.; Xu, Z.; Sugioka, K.; Midorikawa, K. Rapid prototyping of three-dimensional microfluidic mixers in glass by femtosecond laser direct writing. *Lab Chip* **2012**, *12*, 746–749.

15. Talary, M.S.; Burt, J.P.; Rizvi, N.H.; Rumsby, P.T.; Pethig, R. Microfabrication of biofactory-on-a-chip devices using laser ablation technology. *Proc. SPIE* **1999**, *3680*, 572–580.

16. Müller, T.; Gradl, G.; Howitz, S.; Shirley, S.; Schnelle, T.; Fuhr, G. A 3-D microelectrode system for handling and caging single cells and particles. *Biosens. Bioelectron.* **1999**, *14*, 247–256.

17. Cho, C.-H.; Cho, W.; Ahn, Y.; Hwang, S.-Y. PDMS–glass serpentine microchannel chip for time domain PCR with bubble suppression in sample injection. *J. Micromech. Microeng.* **2007**, *17*, 1810–1817.

18. Buehler, S.M.; Stubbe, M.; Gimsa, U.; Baumann, W.; Gimsa, J. A decrease of intracellular ATP is compensated by increased respiration and acidification at sub-lethal parathion concentrations in murine embryonic neuronal cells: Measurements in metabolic cell-culture chips. *Toxicol. Lett.* **2011**, *207*, 182–190.

19. Peters, R.; Oldorf, P. Novel Applications of Ultra-short Pulsed Lasers. *Opt. Photonik* **2012**, *7*, 30–33.

20. Baumann, W.H.; Lehmann, M.; Schwinde, A.; Ehret, R.; Brischwein, M.; Wolf, B. Microelectronic sensor system for microphysiological application on living cells. *Sens. Actuators B Chem.* **1999**, *55*, 77–89.

21. Gimsa, U.; Iglič, A.; Fiedler, S.; Zwanzig, M.; Kralj-Iglič, V.; Jonas, L.; Gimsa, J. Actin is not required for nanotubular protrusions of primary astrocytes grown on metal nano-lawn. *Mol. Membr. Biol.* **2007**, *24*, 243–255.

22. Matschegewski, C.; Staehlke, S.; Birkholz, H.; Lange, R.; Beck, U.; Engel, K.; Nebe, J.B. Automatic Actin Filament Quantification of Osteoblasts and Their Morphometric Analysis on Microtextured Silicon-Titanium Arrays. *Materials* **2012**, *5*, 1176–1195.

23. Schöning, M.J.; Ronkel, F.; Crott, M.; Thust, M.; Schultze, J.W.; Kordos, P.; Lüth, H. Miniaturization of potentiometric sensors using porous silicon microtechnology. *Electrochim. Acta* **1997**, *42*, 3185–3193.

Adaptive Covariance Estimation Method for LiDAR-Aided Multi-Sensor Integrated Navigation Systems

Shifei Liu [1,*], Mohamed Maher Atia [2], Yanbin Gao [1] and Aboelmagd Noureldin [2]

[1] College of Automation, Harbin Engineering University, 145 Nantong St., Nangang District, Harbin 150001, China; E-Mail: gaoyanbin@hrbeu.edu.cn
[2] Department of Electrical and Computer Engineering, Royal Military College of Canada, P.O. Box 17000, Station Forces, Kingston, ON K7K 7B4, Canada; E-Mails: mohamed.maher.atia@gmail.com (M.M.A.); aboelmagd.noureldin@rmc.ca (A.N.)

* Author to whom correspondence should be addressed; E-Mail: shifeiliu@gmail.com

Academic Editors: Naser El-Sheimy and Joost Lötters

Abstract: The accurate estimation of measurements covariance is a fundamental problem in sensors fusion algorithms and is crucial for the proper operation of filtering algorithms. This paper provides an innovative solution for this problem and realizes the proposed solution on a 2D indoor navigation system for unmanned ground vehicles (UGVs) that fuses measurements from a MEMS-grade gyroscope, speed measurements and a light detection and ranging (LiDAR) sensor. A computationally efficient weighted line extraction method is introduced, where the LiDAR intensity measurements are used, such that the random range errors and systematic errors due to surface reflectivity in LiDAR measurements are considered. The vehicle pose change is obtained from LiDAR line feature matching, and the corresponding pose change covariance is also estimated by a weighted least squares-based technique. The estimated LiDAR-based pose changes are applied as periodic updates to the Inertial Navigation System (INS) in an innovative extended Kalman filter (EKF) design. Besides, the influences of the environment geometry layout and line estimation error are discussed. Real experiments in indoor environment are performed to evaluate the proposed algorithm. The results showed the great consistency between the LiDAR-estimated pose change covariance and the true accuracy. Therefore, this leads to a significant improvement in the vehicle's integrated navigation accuracy.

Keywords: LiDAR; MEMS-based INS; UGV; indoor navigation; covariance estimation; multi-sensor integration

---

1. Introduction

The capacity of autonomous localization and navigation for an unmanned ground vehicle (UGV) in an indoor environment is imperative. This enables the vehicle to explore the unknown environment without pre-installation, which also can be applied to visually-impaired persons and first responders. Meanwhile, localization and navigation indoors is challenging due to the absence of Global Positioning System (GPS) signals, which are widely used in outdoor navigation. Generally, an UGV can estimate its position and orientation by integrating the information from inertial sensors, like the gyroscope and accelerometer, which are known as dead reckoning. However, the Inertial Navigation System (INS) suffers from error accumulation. To this end, it is necessary to fuse INS with other complementary sensors to enhance the dead reckoning accuracy in GPS-denied environments. A common sensor available in most UGVs used indoors is light detection and ranging (LiDAR). LiDAR emits a sequence of laser beams at a certain bearing. When hitting the surface of the objects within the scanning range, the laser pulse reflects back. LiDAR estimates distance from the laser scanner to the objects in the environment by recording the time difference between emitted and reflected pulses. The high sampling rate and reliable LiDAR measurements can be used to derive the motion of the vehicle or to interpret the contour of the environment.

Most localization and navigation approaches using LiDAR can be classified into two categories based on the availability of a geometric map [1]. Given the map, the position and orientation of the vehicle can be estimated by matching a scan or features extracted from the scan with the *a priori* map. The main limitation with this map-based localization method is the system's incapability to adjust to spatial layout changes [2]. When the map is unavailable, the vehicle will perform localization by matching scans to track the relative position and orientation changes, thus tracking the vehicle's position and orientation. If a map is required to be generated concurrently, this is called simultaneous localization and mapping (SLAM) [3].

The most common features used in indoor LiDAR-based odometry and SLAM are line features. Since line features universally exist in indoor environments, they are extracted and tracked to perform LiDAR-based indoor localization. The relative position and orientation (pose) changes estimated from LiDAR are fused with INS in a filtering algorithm: the Kalman filter (KF) or the particle filter (PF). However, there are significant challenges involved in this mechanism. Firstly, although line features are dominant in most indoor environments, there are some areas where well-structured line features do not exist. Secondly, it is important to determine how many lines there are in the environment and which point belongs to which line. In addition, tracking the same line between consecutive scans is also extremely important. Thirdly, the LiDAR range and bearing measurements are contaminated with errors due to instrument error, the reflectivity ability of the scanned surface, environmental conditions, like temperature and the atmosphere, *etc.* [4]. When calculating line parameters, the error factors need to be dealt with. Finally, when integrating INS with LiDAR through a filter, the measurement covariance matrix, which indicates how much we can trust the measurements, should be taken into consideration. All of these issues have not been extensively addressed in the literature, especially the measurement covariance estimation issue. Therefore, this paper introduces an adaptive covariance estimation method that can be used in any LiDAR-aided integrated navigation system. The method is applied in an extended Kalman filter (EKF) design that fuses LiDAR, MEMS-grade inertial sensors and the vehicle's odometer observations.

The rest of this paper is organized as follows: Section 2 introduces some of the related works, and Section 3 describes the line extraction methods, including line segmentation, line parameter calculation and line merging. In addition, line feature matching-based pose changes and their adaptive covariance estimation method is described. Then, in Section 4, the integrated system and the filter design are explained in detail. Section 5 presents and discusses the experiment results, while the conclusion is given in Section 6.

2. Related Work

LiDAR is usually mounted on carriers, like an air vehicle, land vehicle or pedestrian, to implement pose estimation, mobile mapping and navigation in both indoor and outdoor environments, unaided or with other sensors. One of the techniques that is commonly adopted in LiDAR-based motion estimation is scan matching. The scan matching methods can be broadly classified into three different categories: feature-based scan matching, point-based scan matching and mathematical property-based scan matching [5]. Feature-based scan matching extracts distinctive geometrical patterns, such as line segments [6–9], corners and jump edges [10,11], lane makers [12] and curvature functions [13], from the LiDAR measurements. Feature-based scan matching is efficient, but it depends on the structure of the environment and has limited ability to cope with unstructured or clustered environments. Point-based scan matching matches scan points directly without extracting any features. Compared with the feature-based scan matching method, it is suitable for more general environments. In addition, since it uses the raw sensor measurements for estimation, it is more accurate, but the computation load increases as well. Specifically, the iterative closest point (ICP) algorithm [14,15] and its variants [16] are the most popular methods dealing with the point-based scan matching problem, due to their simplicity and effectiveness. The core step is to find corresponding point pairs, which represent the same physical place in the environment. This process is difficult to implement, due to interference and noises, and requires iteration to refine the final results. For the mathematical property-based scan matching method, it can be a histogram [17], Hough transformation [18], normal distribution transform [19] or cross-correlation [20].

Despite the limitations of the feature-based scan matching method, indoor localization and navigation can benefit from it, because most indoor environments are commonly well structured, and line features can be easily extracted and used. The line feature extraction process estimates the line parameters from a set of scan points, and it generally includes the following steps:

(1) The first step is preprocessing, which is an optional procedure on the LiDAR measurements. In [10,21], a median filter is used to filter out some of the outliers and noises, while a six order polynomial fitting is used in [22,23] between the measured distance and the true distance to compensate for the systematic error.

(2) The following step is to segment scan points into different groups, each of which includes points belonging to the same line. Generally, there are three segmentation or breakpoint detection methods: one is based on the Euclidean distance between two consecutive scanned points, while the others are KF-based segmentation methods and the fuzzy cluster method [23–25].

(3) Then, the line parameters are computed. The most commonly used line fitting techniques are the Hough transform (HT) [26] and least squares. HT can be used to extract line features by transforming the image pixel into the parameter space and detecting peaks in that space. However, HT has some drawbacks, like the occurrence of multi-peaks due to the discretization of the image and the parameter space, as indicated in [27].

(4) Finally, collinear lines are merged into one line feature making the process more efficient.

A wide range of commonly used line feature extraction methods are compared in [28]. The difference lies in the segmentation method, while total least squares is used to calculate the line parameters and their associated covariance. However, these methods iterate the segmentation and line fitting procedures over all of the scan points. Therefore, they are computationally expensive. In addition, errors in LiDAR measurements are usually neglected. Some earlier papers discussed and dealt with the systematic and random errors in the LiDAR raw measurements. Particularly, the noises in both range and angle measurements are assumed to be zero mean Gaussian random variables [9] and are modeled as the difference between the true value and the noisy measurements. Then, each scan point's influence in the overall line fitting is weighted according to its uncertainty. Similarly, in [11], two sources of error—measurement process noise and quantization error—are modeled. The uncertainty of extracted line features, which depends on the line fitting method and on the range error models, is studies in [29]. By associating and matching the same line features in different scans, the vehicle relative motion change between the two positions where scans are taken can be estimated.

A lot of the work on LiDAR measurements and the feature uncertainty discussion do not cover the further analysis of the integration with other sensors, like the work in [8,30]. In [8], the modified incremental split and merge algorithm is proposed for line extraction. Besides, the standard deviation of the perpendicular distance from the scan points to the extracted line feature is calculated to evaluate the quality of the extracted lines and is used in the calculation of the covariance of LiDAR observations (perpendicular distance changes in this case). Meanwhile, INS outputs are used to aid LiDAR feature prediction and matching, tilt compensation and laser-based navigation solution computation when not enough features are extracted.

Different from existing methods, in this paper, an efficient LiDAR-based pose change estimation method supported by a novel adaptive measurement covariance estimation technique is introduced. Furthermore, an EKF design that fuses LiDAR, inertial sensors and the odometer is given to provide a robust integrated navigation system for UGVs. The main contributions of this paper are listed below:

(1) The covariance of LiDAR-derived relative pose change is estimated. This is crucial when the LiDAR-derived relative pose changes are to be fused with information from other sensors. The proposed adaptive covariance estimation method can be applied in any LiDAR-based integrated navigation system.

(2) The LiDAR intensity measurements are used to weight the influence of each scanned point on the line parameters estimation.

(3) The influences of the geometric layout of the environment (especially the long corridor) and line feature extraction error are addressed.

3. Line Extraction Method

3.1. Segmentation

3.1.1. Breakpoint Detection

The purpose of segmentation is to identify line features in the environment and to ignore meaningless scan points, which are likely to be scattered and unorganized objects, such as desks and chair legs. The first step of segmentation is the breakpoint detection. The breakpoint is defined as the discontinuity in the range measurements, and the goal of breakpoint detection is to roughly group scan points into subgroups, which are potential line features. Given the fact that the same features representing the same landmarks in the environment are almost consecutive points [7], therefore, the LiDAR range measurements for the same feature should change monotonously and smoothly. Once there is a significant change in the range measurements between two consecutive scan points, a breakpoint is declared. The breakpoint detection criterion is given as:

$$\left| r_{m+1} - r_m \right| \geq r_{threshold} \tag{1}$$

where r_m and r_{m+1} are two consecutive range measurements for the m-th and $(m + 1)$-th scan points and $r_{threshold}$ is the breakpoint detection threshold.

3.1.2. Corner Detection

The breakpoint detection criterion does not apply to corner detection, since a corner is the intersection of two line features, and the range measurements transit smoothly. Figure 1 shows one of the scan scenes and the corresponding range measurements.

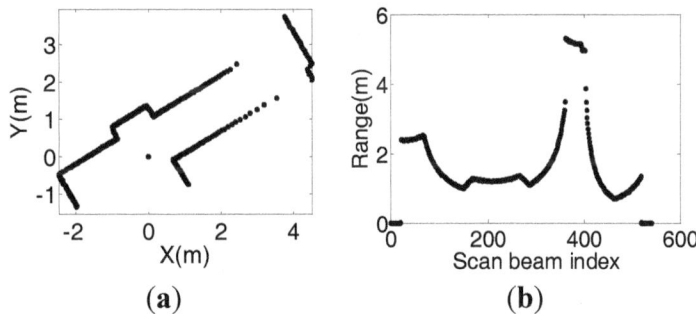

Figure 1. Corner measurements. (**a**) One of the corner scan scenes. (**b**) The corresponding range measurements.

From the scan scene, there are ten lines in total. However, when applying the breakpoint detection criterion and ignoring short segments containing less than N points, which tend to be scattered objects and are susceptible to noises and interferes, only three lines will be extracted, since corners cannot be identified. Here, N is the minimum number of points on the line and is set according to the distance from the scanned point to the sensor location.

Figure 2 demonstrates one of the scan scenes where three corners exist and are illustrated with red squares. After analyzing the pattern of corners, we realize that for M consecutive scan points, if a corner exists, the corresponding range measurement of the corner point is likely to be either the maximum or the minimum value among the M range measurements, as shown in Figure 2.

However, since LiDAR range measurements suffer from systematic and random noises, only detecting the maximum or minimum range measurement among M scan points can result in the detection of a spurious corner. To increase the robustness of the proposed line extraction method, the following criterion is added as a complementary to detect corners after the extremum r_m among M range measurements is detected:

$$\sum_{j=m-M/2}^{m+M/2} \left| r_m - r_j \right| \geq r_{corner\_threshold} \tag{2}$$

where $r_{corner\_threshold}$ is the corner detection threshold. It is important to note for points on the same line feature, especially right beside the sensor, there will be points satisfying the second criterion. Therefore, the line feature will be detected as being more than one. However, this issue will be solved properly in the line merging process. Comparing our line extraction algorithm with the previous works, the corner problem is addressed, and iteration is not required.

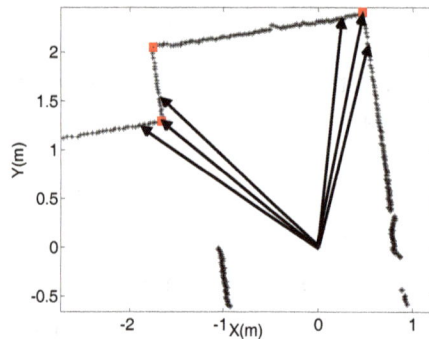

Figure 2. Corner detection.

3.2. Line Parameters Calculation

After line segmentation, the weighted least squares method is used in our work to calculate the two line parameters in polar coordinate. The two parameters are the perpendicular distance from the origin of the vehicle body frame to the extracted line ρ and the angle between the x-axis of the body frame and the norm of the extracted line α, as shown in Figure 3.

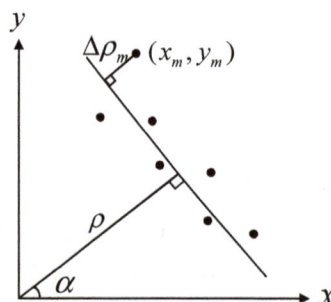

Figure 3. Line feature represented in the vehicle body frame.

3.2.1. Scan Points' Weight Calculation

Before applying weighted least squares to the line parameters estimation, the weight of each scan point is computed. Assuming that the diffuse component of the laser reflection is Lambertian, the following relation can be modeled [31]:

$$P_{return} \propto \frac{p \cos \beta}{r^2} \tag{3}$$

where P_{return} is the intensity of the returned laser pulse, p is the surface reflectance ranging from zero to one, β is the angle of incidence of the laser beam with the surface and r is the distance from the detected object to the sensor. As can be seen from this relation, the intensity is linked to the surface reflectivity, which will result in systematic error in range measurements [4]. Hence, the intensity will be used to weight the noise level in each LiDAR scanned point.

Regarding consecutive points on the same line segment, they share the same surface reflectance, and the difference between the corresponding range measurements is very small. The only significant difference is the incidence angle of the laser beam with the surface.

The reflected intensity of the m-th laser beam can be given as:

$$P_m \propto \frac{p \cos(\theta_m - \alpha)}{r_m} \tag{4}$$

where θ_m is the bearing of the m-th laser beam. Assume that consecutive points on the same line segment have the same range measurements and differential intensities; we can get:

$$\Delta P_{m,m+1} \propto \frac{p}{r_m} \left[\cos(\theta_{m+1} - \alpha) - \cos(\theta_m - \alpha) \right] \tag{5}$$

Apply the Taylor series to approximate the cosine function in the above equation, and only keep the first two terms; this can be rewritten as:

$$\Delta P_{m,m+1} \propto \frac{p}{2r_m}(-\Delta\theta)(2\theta_{m+1} - \Delta\theta - 2\alpha) \tag{6}$$

where $\Delta\theta$ is the bearing difference of two consecutive laser beams, also called the angle resolution. If we differentiate the intensity changes, the following relation can be derived:

$$\nabla \Delta P_{m-1,m+1} \propto -\frac{p}{r_m}(\Delta\theta)^2 \tag{7}$$

The angle resolution in our work is 0.5 degrees. Transformed to radians and squared, the numerator is at a magnitude of 10^{-6}, which indicates that the double difference of the consecutive points' intensity measurements is approximately zero. Therefore, for points on the same line segment, the double difference of their intensities should be very close to zero. Significant deviation from zero is mainly caused by errors and will be used as the weight for the scan points. The more significant the deviation is, the lower the weight is.

3.2.2. Line Parameters Estimation

After deriving the weight for each scan point, the line parameters ρ and α are estimated. Line parameters estimation is the problem of minimizing the sum of the squared normal distance from each point to the detected line [6]. The cost function is given as:

$$\min \sum w_m \left(\rho - x_m \cos\alpha - y_m \sin\alpha \right)^2 \tag{8}$$

Here, (x_m, y_m) are the coordinates of the m-th scan point in the Cartesian frame. The least-squares-based solution for the above cost function is given as follows [6]:

$$\alpha = \frac{1}{2}\arctan\left(\frac{-2S_{xy}}{S_{yy} - S_{xx}}\right) \tag{9}$$

$$\rho = \bar{x}\cos\alpha + \bar{y}\sin\alpha \tag{10}$$

where $S_{xy} = \sum w_m(x_m - \bar{x})(y_m - \bar{y})$, $S_{yy} = \sum w_m(y_m - \bar{y})^2$, $S_{xx} = \sum w_m(x_m - \bar{x})^2$, $\bar{x} = \frac{1}{\sum w_m}\sum w_m x_m$, $\bar{y} = \frac{1}{\sum w_m}\sum w_m y_m$.

3.2.3. Line Quality Calculation

To evaluate the line extraction error, we define a parameter, line quality. When the residual of the cost function displayed in Equation (8) is small, we assume that the line extraction error is small and that the quality of the line is good, and *vice versa*. Before giving the definition for the line quality parameter, we first define the perpendicular distance from the m-th scan point to the line as $\Delta\rho_m$, as demonstrated in Figure 3. It can be depicted as:

$$\Delta\rho_m = \rho - x_m \cos\alpha - y_m \sin\alpha \tag{11}$$

The quality of the k-th line feature q_k is thereafter defined as the covariance of the perpendicular distance deviation of all of the n points on the line, denoted as:

$$q_k = \text{cov}(\Delta\rho_1, ..., \Delta\rho_n) \tag{12}$$

When estimating pose change and the associated covariance in the following line feature matching process, a line feature with a big extraction error will be given less weight, and a line feature of good quality will have a big weight.

3.3. Line Merging

There are cases when a line feature in the real world is partly blocked or interrupted. To this end, more than one line representing the same line feature will be extracted. Merging these lines can gain more reliable line features and enable the line feature matching process to be more efficient. The line merging criterion is to compare the parameters of the detected line features, since the line parameters of the same line feature should be close. If the differences between the line parameters are small, the lines are merged, and new line parameters are recomputed.

3.4. Pose Change and Covariance Estimation

After line features are extracted and merged, the same line features occurring in the two consecutive scans are matched to derive the relative position and azimuth (pose) changes. The line feature matching is implemented by checking each line feature in the previous scan against the line features in the next scan. If the relative azimuth and perpendicular distance changes are close within a threshold, a matched line feature pair is declared. To illustrate how vehicle pose change is derived from line feature matching, Figure 4 shows the same line feature (shown in red) in the body frame at epoch i and $i + 1$. The relative position changes along both axes are defined as Δx and Δy while the azimuth change is defined as $\Delta A_{i,i+1}$.

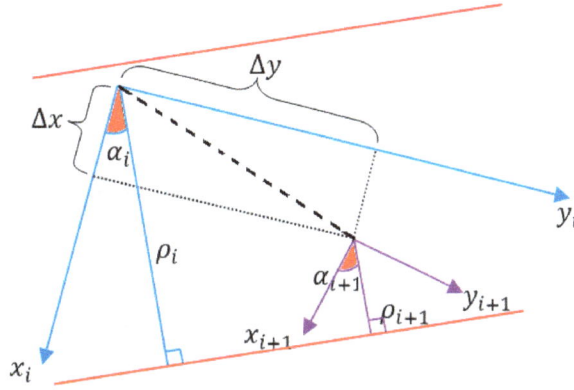

Figure 4. The vehicle pose change over two consecutive scans.

3.4.1. Position Change and Covariance Estimation

The relation between perpendicular distance change and relative position change is given as below:

$$\Delta \rho_{i,i+1} = \rho_i - \rho_{i+1} = \Delta x \cos \alpha_i + \Delta y \sin \alpha_i \tag{13}$$

Assume that k line features are extracted and matched for two consecutive scans; Equation (13) can be rewritten as:

$$\begin{bmatrix} \Delta \rho_{i,i+1,1} \\ ... \\ \Delta \rho_{i,i+1,k} \end{bmatrix} = H_\rho \begin{bmatrix} \Delta x \\ \Delta y \end{bmatrix} = \begin{bmatrix} \cos \alpha_{i,1} & \sin \alpha_{i,1} \\ ... & ... \\ \cos \alpha_{i,k} & \sin \alpha_{i,k} \end{bmatrix} \begin{bmatrix} \Delta x \\ \Delta y \end{bmatrix} \tag{14}$$

Here, $\Delta \rho_{i,i+1,j}$ is the perpendicular distance change of the j-th pair of the matched line feature between epoch i and $i + 1$. H_ρ is the design matrix with elements that are cosine and sine terms of $\alpha_{i,j}$, while $\alpha_{i,j}$ is the angle parameter of the j-th pair of the matched line feature in epoch i, $j = 1, ..., k$. Applying weighted least squares, the solution of the displacement vector can be given as:

$$\begin{bmatrix} \Delta x \\ \Delta y \end{bmatrix} = (H_\rho^T W H_\rho)^{-1} H_\rho^T W \begin{bmatrix} \Delta \rho_{i,i+1,1} \\ ... \\ \Delta \rho_{i,i+1,k} \end{bmatrix} \tag{15}$$

where W is the diagonal matrix representing the weight of the observation vector, which is the perpendicular distance change vector. The j-th element ($j = 1, ..., k$) of W is denoted as:

$$W_j = \frac{1}{q_{i,j} + q_{i+1,j}} \tag{16}$$

where $q_{i,j}$ and $q_{i+1,j}$ are derived from line quality calculation Equation (12). By defining in this way, the weight of the matched line feature pair is inversely proportional to the line extraction errors of the matched lines. The covariance matrix of the displacement vector can then be given as follows:

$$C_{\Delta x, \Delta y} = (H_\rho^T W H_\rho)^{-1} \sigma_\rho^2 \tag{17}$$

where σ_ρ^2 is the perpendicular distance change error variance.

3.4.2. Azimuth Change and Covariance Estimation

The relation between line feature angle change and vehicle azimuth change is straightforward, which can be represented as:

$$\Delta A_{i,i+1} = \alpha_{i+1} - \alpha_i \tag{18}$$

Similarly, given k observations, the azimuth change $\Delta A_{i,i+1}$ can be expressed as:

$$\begin{bmatrix} \Delta \alpha_{i,i+1,1} \\ ... \\ \Delta \alpha_{i,i+1,k} \end{bmatrix} = H_\alpha \Delta A_{i,i+1} = \begin{bmatrix} 1 & ... & 1 \end{bmatrix}^T \Delta A_{i,i+1} \tag{19}$$

where $\Delta \alpha_{i,i+1,j}$ is the line feature angle change of the j-th pair of the matched line feature between epoch i and $i + 1$, $j = 1, ..., k$. H_α is a vector with elements of one. Applying weighted least squares, the solution of the azimuth change can be given as:

$$\Delta A_{i,i+1} = (H_\alpha^T W H_\alpha)^{-1} H_\alpha^T W \begin{bmatrix} \Delta \alpha_{i,i+1,1} \\ ... \\ \Delta \alpha_{i,i+1,k} \end{bmatrix} \tag{20}$$

where W is the diagonal matrix representing the weight of each pair of matched line feature angle parameter changes between continuous scans. It is defined by Equation (16).

For the covariance of the azimuth change, this is given as:

$$C_{\Delta A} = (H_\alpha^T W H_\alpha)^{-1} \sigma_\alpha^2 \tag{21}$$

where σ_α^2 is the line feature angle change error variance.

3.4.3. Singularity Issue

As pointed out by [8], the transformation from line extraction errors into position error is determined by the line geometry. More specifically, when only one line feature or only parallel line features are matched, the relative position change solutions of the weighted least squares will be singular, since the rank of the design matrix will be less than the number of the expected variables, which is two in this case. This is an issue occurring especially in long corridors. On the contrary, when at least two perpendicular lines are matched, the singularity issue will not exist, and the relative position change can be accurately estimated. Therefore, it is essential that the estimated position change covariance can reveal the influence of the line geometry. Besides, the limitation of LiDAR navigating in long corridors also makes the integration of LiDAR and INS of great significance. When the singularity issue occurs, the inverse of $H_\rho^T W H_\rho$ in Equations (15) and (17) will be significantly large. Therefore, the estimated position change will be erroneous, and the associated covariance estimation will be large. In the fusion with the gyroscope and odometer through the filter, the erroneous updates will be reflected by the estimated observation covariance, and consequently, the filter will trust the prediction from the system model more than the LiDAR-derived measurements in this case. It is important to mention that the singularity issue does not exist in the azimuth change estimation.

4. The Integrated Solution and Filter Design

In this section, the integrated solution is described. The main motion model of the proposed system is an INS supported by odometer measurements. Inertial sensors can provide angular velocity and acceleration, while the odometer can provide linear velocity. They are independent of the operating environment, but suffer from error accumulation. On the other side, pose tracking using LiDAR line feature-based scan matching has long-term consistent accuracy, but it is susceptible to the environment's structure. To achieve better performance than using any sensor unaided, the three sensors are integrated through an EKF. One innovation of our filter design is that only one vertical gyroscope is used without the accelerometer, given the assumption that land-based vehicles, especially UGVs, mostly move in horizontal planes [32,33]. The benefits are the lower cost and less complexity in mechanization. More importantly, velocity is derived from the odometer instead of the integration results of the accelerometer outputs; hence, the error in velocity is reduced. The block diagram of the multi-sensor integrated navigation system is demonstrated in Figure 5.

More specifically, a single-axis gyroscope vertically aligned with the vehicle body frame is used to detect the rotation rate of the body frame with respect to the inertial frame, from which the azimuth can be yielded. The odometer can output velocity along the direction where the vehicle moves forward. By integrating the single-axis gyroscope and odometer, the vehicle motion change from dead reckoning can be derived. Meanwhile, the pose change and covariance can be derived from LiDAR based on the proposed algorithm. Due to the nonlinearity of the system, the EKF is adopted to fuse the information from different sensors. Then, the gyroscope bias is accurately estimated, and the navigation solutions are corrected.

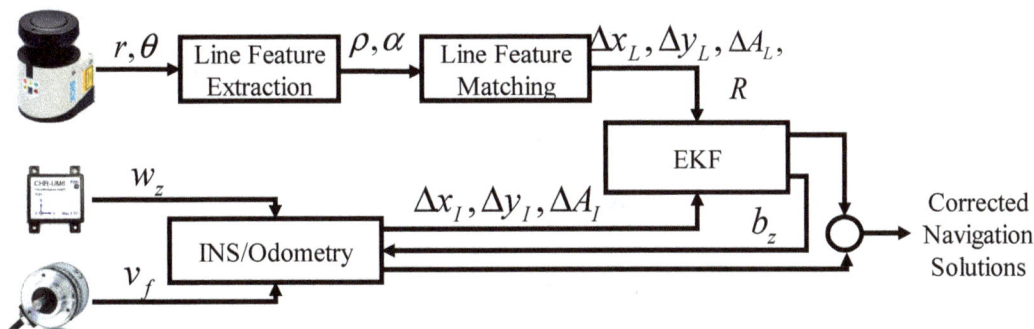

Figure 5. The block diagram of the multi-sensor integrated navigation system.

4.1. System Model

The error state vector for the filter design is defined as:

$$\delta x = \begin{bmatrix} \delta\Delta x & \delta\Delta y & \delta v_f & \delta v_x & \delta v_y & \delta\Delta A & \delta a_{od} & \delta b_z \end{bmatrix}^T$$

where the variables are defined as below: $\delta\Delta x$ is the error in the relative displacement between scans along the x-axis in the vehicle body frame; $\delta\Delta y$ is the error in the relative displacement between scans along the y-axis in the vehicle body frame; δv_f is the error in the forward speed derived from the odometer; δv_x is the error in the forward velocity projecting along the x-axis in the vehicle body frame; δv_y is the error in the forward velocity projecting along the y-axis in the vehicle body frame; $\delta\Delta A$ is the error in the azimuth change between scans; δa_{od} is the error in acceleration derived from the odometer measurements; δb_z is the error in the gyroscope bias.

The system model is derived from INS/odometry mechanization. Specifically, the azimuth change between two scans can be calculated using gyroscope measurements. It is given as follows:

$$\Delta A_I = -(w_z - b_z)T \tag{22}$$

where ΔA_I is the azimuth change obtained from INS, w_z is the gyroscope measurement, b_z is the gyroscope bias and T is the time interval between two scans. Based on the azimuth change, forward velocity can be projected to the vehicle body frame:

$$v_x = v_f \sin(\Delta A_I) \tag{23}$$

$$v_y = v_f \cos(\Delta A_I) \tag{24}$$

Therefore, the relative position change between two scans can be denoted as:

$$\Delta x_I = v_x T \tag{25}$$

$$\Delta y_I = v_y T \tag{26}$$

By applying Taylor expansion to the INS/odometry dynamic system given in Equations (22)–(26) and considering only the first order term, the linearized dynamic system error model is given as below [34,35]:

$$\delta\Delta\dot{x} = \delta v_x \tag{27}$$

$$\delta\Delta\dot{y} = \delta v_y \tag{28}$$

$$\delta\dot{v}_f = \delta a_{od} \tag{29}$$

$$\delta\dot{v}_x = \sin(\Delta A)\delta a_{od} + \cos(\Delta A)(w_z - b_z)\delta v_f - v_f\cos(\Delta A)\delta b_z + \left[a_{od}\cos(\Delta A) - v_f\sin(\Delta A)(w_z - b_z)\right]\delta\Delta A \tag{30}$$

$$\delta\dot{v}_y = \cos(\Delta A)\delta a_{od} - \sin(\Delta A)(w_z - b_z)\delta v_f + v_f\sin(\Delta A)\delta b_z - \left[a_{od}\sin(\Delta A) + v_f\cos(\Delta A)(w_z - b_z)\right]\delta\Delta A \tag{31}$$

$$\delta\Delta\dot{A} = -\delta b_z \tag{32}$$

$$\delta\dot{a}_{od} = -\gamma_{od}\delta a_{od} + \sqrt{2\gamma_{od}\sigma_{od}^2}\,w \tag{33}$$

$$\delta\dot{b}_z = -\beta_z\delta b_z + \sqrt{2\beta_z\sigma_z^2}\,w \tag{34}$$

Here, both random errors in acceleration derived from odometer and gyroscope measurements are modeled as first order Gauss–Markov processes. γ_{od} and β_z are the reciprocal of the correlation time constants of the random process, while σ_{od} and σ_z are the standard deviations associated with the odometer and gyroscope measurements, respectively.

4.2. Measurement Model

For the measurement model, the pose changes estimated from the line feature scan matching are used as filter observations, and the observation covariance matrix is also adaptively calculated by the proposed technique. The advantages of this filter design are two-fold: (1) the line geometry influence is taken into account when estimating the observation covariance; thus, it can precisely reflect the error level in the observations; (2) the tuning of the observation covariance matrix is avoided.

The difference of vehicle pose change between scans obtained from INS/odometry and LiDAR is taken as the observation vector in the measurement model. The observation vector is defined by z and can be given as:

$$z = \begin{bmatrix} \Delta x_L - \Delta x_I \\ \Delta y_L - \Delta y_I \\ \Delta A_L - \Delta A_I \end{bmatrix} \tag{35}$$

The subscript L and I denote measurements from LiDAR and INS, respectively. Therefore, the design matrix for the measurement model can be represented as:

$$H = \begin{bmatrix} 1 & 0 & 0 & 0 & 0 & 0 & 0 & 0 \\ 0 & 1 & 0 & 0 & 0 & 0 & 0 & 0 \\ 0 & 0 & 0 & 0 & 0 & 1 & 0 & 0 \end{bmatrix} \tag{36}$$

For the observation covariance matrix R, it is assumed to be a diagonal matrix without considering the correlation between relative displacement and azimuth change. It is given as:

$$R = diag(C_{\Delta x, \Delta y}, C_{\Delta A}) \tag{37}$$

Here, $C_{\Delta x, \Delta y}$ and $C_{\Delta A}$ are calculated using Equations (17) and (21), respectively.

After establishing the system model and measurement model, EKF equations are used to implement the prediction and correction. The prediction and correction are performed in the body frame and then transformed into the navigation frame to provide corrected navigation outputs.

5. Experiment Results and Analysis

To evaluate the proposed algorithm, real experiments have been performed in an indoor office building with a UGV called Husky A200 from Clearpath Robotics Inc. (Waterloo, Canada), shown in Figure 6. This UGV is wirelessly controlled, and the platform is equipped with a CHR-UM6 MEMS-grade inertial measurement unit (IMU) [36], a SICK laser scanner LMS111 (SICK, Waldkirch, Germany) [37] and a quadrature encoder. The specifications of the laser scanner are shown in Table 1.

The experiment is conducted on the second floor of the building, with the UGV moving in the corridor during the whole process. The spatial layout of the floor is an approximate 70 m by 40 m rectangle, shown in Figure 7.

As can be seen from the floor plan, the corridor length sometimes can exceed the maximum scanning range of the LiDAR, which is 20 m in our experiments. This may result in the singularity problem. From the earlier discussion, it can be seen that a poor geometric layout can cause the singularity issue, and the estimated covariance will be large. Meanwhile, due to the proper weighting scheme for the line feature in the covariance estimation process, a large line extraction error will not jeopardize the estimation accuracy of the relative position change and associated covariance. To address the influences of geometry layout and line quality on the relative position change and the covariance estimation, two cases are discussed. The LiDAR scans under these two situations are shown in Figures 8 and 9 with extracted lines marked. The green square represents the beginning of the line, and the red diamond means the end of the line. In both figures, the graph on the left shows the previous scan, while the graph on the right is the scan ten epochs later. The ten-epoch interval is to guarantee distinguishable changes can accumulate and be estimated.

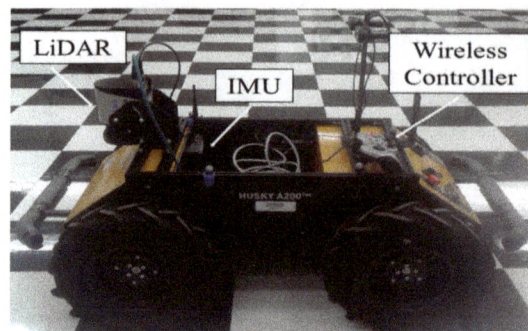

Figure 6. Husky A200.

Table 1. SICK LMS111 specifications.

Parameter	Value
Statistical Error (mm)	±12
Angular Resolution (°)	0.5
Maximum Measurement Range (m)	20
Scanning Range (°)	270
Scanning Frequency (Hz)	50

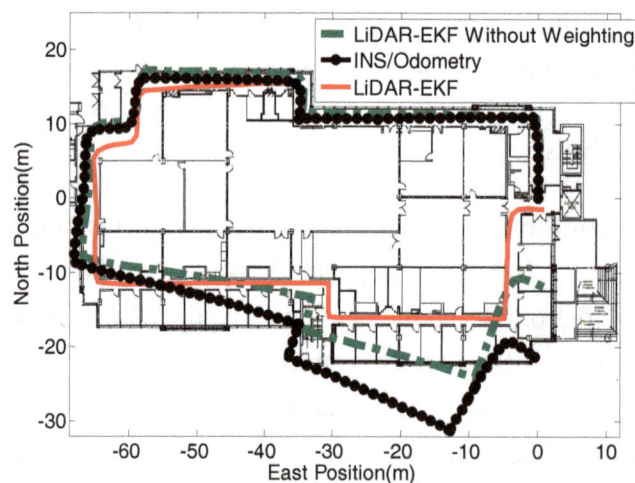

Figure 7. LiDAR-aided and INS/Odometry trajectories.

Figure 8 illustrates an example of when the UGV is located in the long corridor. In this case, only two parallel walls are detected, and two parallel line features will be matched between the two scans. The geometry layout will cause the singularity in the relative position change estimation. However, since the scan points on the extracted lines are well distributed, the line quality is good.

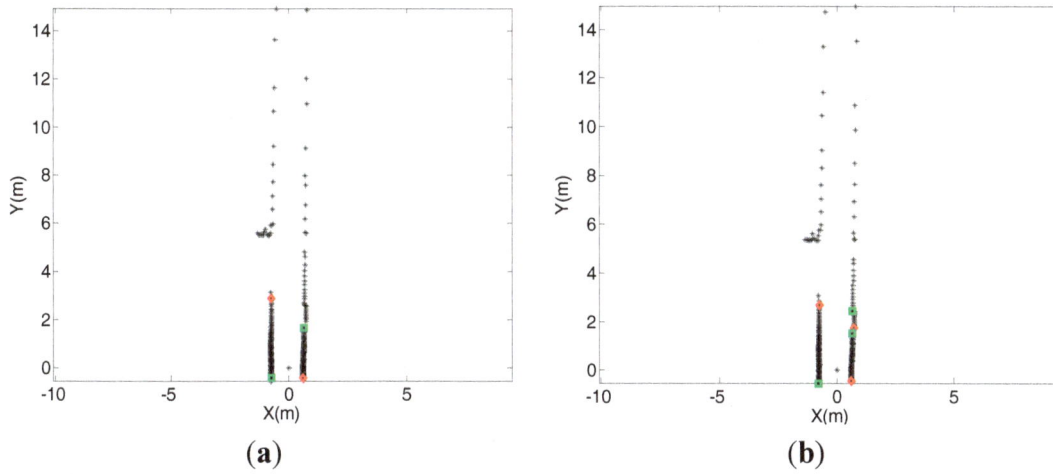

(a)   (b)

Figure 8. UGV moves in the long corridor. (**a**) The first scan. (**b**) The second scan.

Figure 9 demonstrates another example when the UGV comes to a corner. Four lines are extracted in the first scan, while five lines are extracted in the second scan. In the line matching step, the first three lines in both scans are matched successfully, while the fourth line in the first scan is matched with the fifth line in the second scan. For the four pairs of matched lines, the line extraction errors of the first two pairs are small. The third pair's line extraction error will increase, while the last pair's line extraction error will be the most significant. As indicated by Equation (16), the bigger the line extraction error is, the lower the weight that will be given to the corresponding line pair in the line feature scan matching-based pose change and covariance estimation process. To this end, the mismatched fourth line pair will have the least weight in position change estimation.

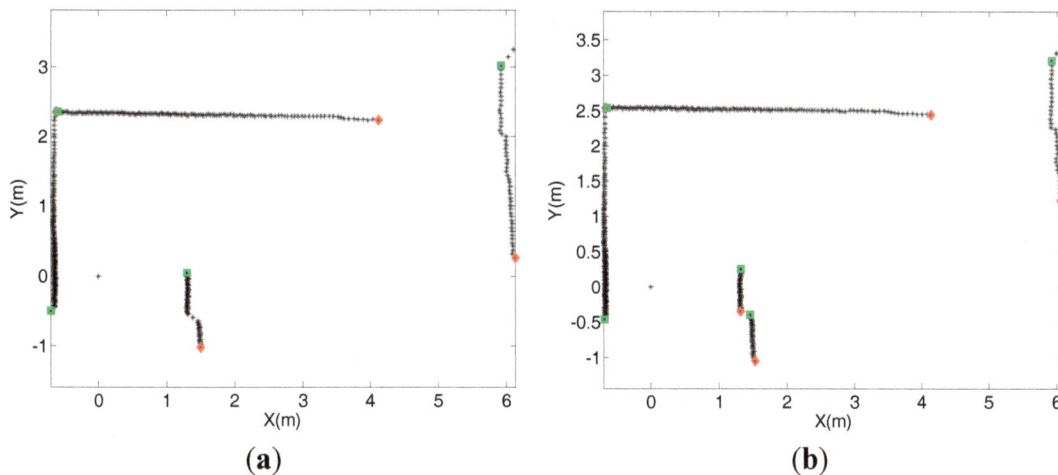

(a)   (b)

Figure 9. The UGV arrives at a corner. (**a**) The first scan. (**b**) The second scan.

The relative displacement and covariance estimation results for the above two examples are shown in Table 2. Under both situations, the UGV mainly moves along the vertical direction of the body frame. The maximum speed of the vehicle is 1 m/s while ten epoch time interval is 0.2 s. Therefore, the results of example 2 are reasonable while the results of example 1 are quite erroneous. This leads to the conclusion that the geometry layout of the extracted lines has significant impact on the estimation of the displacement. On the contrary, since the line feature is weighted by its quality in displacement estimation, line quality has minor effect on the results. However, the covariance of the displacement estimated from the proposed algorithm can reflect the error level in position change precisely. When the displacement is propagated to the filter, the observation can be evaluated according to its covariance and accurate error states estimation can be achieved.

Table 2. SICK LMS111 specifications.

Example	Δx (m)	Δy (m)	Cov (Δx)	Cov (Δy)
Example 1	0.3801	30.1185	0.0041	20.6805
Example 2	0.0602	0.2057	0.0001025	0.0001956

Figure 10 shows a portion of the relative position change in the vehicle forward direction from LiDAR and INS along with the estimated covariance. From the figure, it can be clearly seen that when the LiDAR-derived position change is accurate, the estimated covariance is very small. On the contrary, when the LiDAR-derived position change is significantly jeopardized, the estimated covariance is fairly large. This demonstrates the consistency and the accuracy of the proposed covariance estimation method.

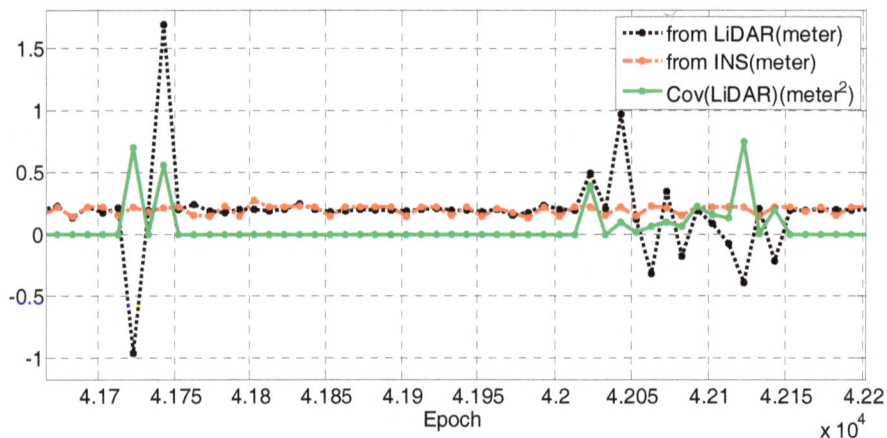

Figure 10. Relative position change in the vehicle forward direction from LiDAR and INS and the estimated covariance.

The trajectories obtained from the unaided INS/odometry dead reckoning system after initial gyroscope bias compensation and the LiDAR-aided integrated system are demonstrated on the floor plan. Besides, the filtered trajectory without weighting the scanned points in the line parameter estimation and without weighting the matched line pairs in the pose change and the associated covariance estimation are also shown in Figure 7. It is important to note that, even without the weighting scheme, the observation covariance still adaptively changes instead of being set to a fixed value. However, due to the inaccurate pose change and associated covariance estimation, the error in the navigation solutions accumulates gradually. Compared with the unaided and unweighted navigation solutions, the filtered navigation solutions show a significant improvement in accuracy. This means that with the corrections from the LiDAR and observation covariance matrix from the proposed algorithm, the gyroscope bias is precisely estimated (illustrated in Figure 11), which directly reflects the improved accuracy of the integrated solutions (the red curve in Figure 7).

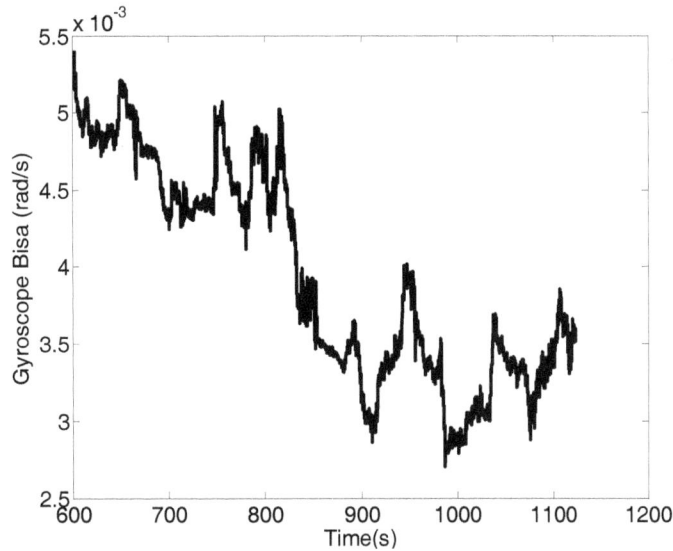

Figure 11. Gyroscope bias estimation results.

6. Conclusions

To overcome the challenge of indoor navigation and to address the limitations in the existing literature, this paper proposed a multi-sensor integrated navigation system that integrates a low-cost MEMS-grade inertial sensor gyroscope, odometer and ranging sensor LiDAR through an EKF. The paper introduced an adaptive covariance estimation method for LiDAR-estimated pose changes. The proposed EKF scheme benefited from the long-term accuracy of LiDAR and the self-contained dead reckoning INS to enhance the overall navigation solutions. The observations derived from LiDAR after line extraction and matching are relative position change and azimuth change. In the line extraction process, the LiDAR raw range and intensity measurements are used, and each scan point is weighted by the noise level in its range measurement. In the line matching step, the singularity issue and the line extraction error are addressed. Besides, the LiDAR observation covariance matrix is also estimated in the line matching step using weighted least squares. In the filter design, LiDAR observations and the observation covariance matrix are propagated to the measurement model. The real experiment results in an indoor building show that the observation covariance can precisely reflect the error level in the observations, and the error accumulation of the vehicle position is significantly reduced compared with the unaided solution.

Author Contributions

Shifei Liu and Mohamed Maher Atia performed the experiments and analyzed the data. Shifei Liu wrote the paper, and Mohamed Maher Atia offered plenty of comments. Yanbin Gao and Aboelmagd Noureldin provided much support and useful discussion.

Conflicts of Interest

The authors declare no conflict of interest.

References

1.   Gutmann, J.-S.; Schlegel, C. Amos: Comparison of scan matching approaches for self-localization in indoor environments. In Proceedings of the First Euromicro Workshop on Advanced Mobile Robot, Kaiserslautern, Germany, 9–11 October 1996; pp. 61–67.

2.   Hesch, J.A.; Mirzaei, F.M.; Mariottini, G.L.; Roumeliotis, S.I. A laser-aided inertial naviagtion system (L-INS) for human localization in unknown indoor environments. In Proceedings of the 2010 IEEE International Conference on Robotics and Automation (ICRA), Anchorage, AK, USA, 3–7 May 2010.

3.   Kohlbrecher, S.; Von Stryk, O.; Meyer, J.; Klingauf, U. A flexible and scalable slam system with full 3D motion estimation. In Proceedings of the 2011 IEEE International Symposium on Safety, Security, and Rescue Robotics (SSRR), 1–5 November 2011; pp. 155–160.

4.   Boehler, W.; Bordas Vicent, M.; Marbs, A. Investigating laser scanner accuracy. *Int. Arch. Photogramm. Remote Sens. Spat. Inf. Sci.* **2003**, *34*, 696–701.

5.   Martínez, J.L.; González, J.; Morales, J.; Mandow, A.; García-Cerezo, A.J. Mobile robot motion estimation by 2D scan matching with genetic and iterative closest point algorithms. *J. Field Robot.* **2006**, *23*, 21–34.

6.   Garulli, A.; Giannitrapani, A.; Rossi, A.; Vicino, A. Mobile robot SLAM for line-based environment representation. In Proceedings of the 44th IEEE Conference on Decision and Control, 2005 and 2005 European Control Conference (CDC-ECC'05), Seville, Spain, 12–15 December 2005; pp. 2041–2046.

7.   Arras, K.O.; Siegwart, R.Y. *Feature Extraction and Scene Interpretation for Map-Based Navigation and Map Building*; Intelligent Systems & Advanced Manufacturing, International Society for Optics and Photonics: Pittsburgh, PA, USA, 1998; pp. 42–53.

8.   Soloviev, A.; Bates, D.; van Graas, F. Tight coupling of laser scanner and inertial measurements for a fully autonomous relative navigation solution. *Navigation* **2007**, *54*, 189–205.

9.   Pfister, S.T.; Roumeliotis, S.I.; Burdick, J.W. Weighted line fitting algorithms for mobile robot map building and efficient data representation. In Proceedings of the IEEE International Conference on Robotics and Automation (2003ICRA'03), 14–19 September 2003; pp. 1304–1311.

10. Lingemann, K.; Nüchter, A.; Hertzberg, J.; Surmann, H. High-speed laser localization for mobile robots. *Robot. Auton. Syst.* **2005**, *51*, 275–296.

11. Aghamohammadi, A.A.; Taghirad, H.D.; Tamjidi, A.H.; Mihankhah, E. Feature-based range scan matching for accurate and high speed mobile robot localization. In Proceedings of the 3rd European Conference on Mobile Robots, Freiburg, Germany, 19–21 September 2007.

12. Kammel, S.; Pitzer, B. Lidar-based lane marker detection and mapping. In Proceedings of the 2008 IEEE Intelligent Vehicles Symposium, Eindhoven, The Netherlands, 4–6 June 2008; pp. 1137–1142.

13. Núñez, P.; Vazquez-Martin, R.; Bandera, A.; Sandoval, F. Fast laser scan matching approach based on adaptive curvature estimation for mobile robots. *Robotica* **2009**, *27*, 469–479.

14. Shen, S.; Michael, N.; Kumar, V. Autonomous multi-floor indoor navigation with a computationally constrained MAV. In Proceedings of the 2011 IEEE International Conference on Robotics and automation (ICRA), Shanghai, China, 9–13 May 2011; pp. 20–25.

15. Segal, A.; Haehnel, D.; Thrun, S. Generalized-ICP. In Proceedings of the Robotics: Science and Systems 2009 Conference, University of Washington, Seattle, WA, USA, 28 June–1 July 2009.

16. Censi, A. An ICP variant using a point-to-line metric. In Proceedings of the 2008 IEEE International Conference on Robotics and Automation (ICRA 2008), Pasadena, CA, USA, 19–23 May 2008; pp. 19–25.

17. Weiß, G.; Puttkamer, E. A map based on laserscans without geometric interpretation. *Intell. Auton. Syst.* **1995**, *4*, 403–407.

18. Censi, A.; Iocchi, L.; Grisetti, G. Scan matching in the Hough domain. In Proceedings of the 2005 IEEE International Conference on Robotics and Automation (ICRA 2005), Barcelona, Spain, 18–22 April 2005; pp. 2739–2744.

19. Burguera, A.; González, Y.; Oliver, G. On the use of likelihood fields to perform sonar scan matching localization. *Auton. Robot.* **2009**, *26*, 203–222.

20. Olson, E.B. Real-time correlative scan matching. In Proceedings of the IEEE International Conference on Robotics and Automation (ICRA'09), Kobe, Japan, 12–17 May 2009; pp. 4387–4393.

21. Diosi, A.; Kleeman, L. Laser scan matching in polar coordinates with application to SLAM. In Proceedings of the 2005 IEEE/RSJ International Conference on Intelligent Robots and Systems (IROS 2005), Edmonton, AB, Canada, 2–6 August 2005; pp. 3317–3322.

22. Núñez, P.; Vázquez-Martín, R.; Del Toro, J.; Bandera, A.; Sandoval, F. Natural landmark extraction for mobile robot navigation based on an adaptive curvature estimation. *Robot. Auton. Syst.* **2008**, *56*, 247–264.

23. Borges, G.A.; Aldon, M.-J. Line extraction in 2D range images for mobile robotics. *J. Intell. Robot. Syst.* **2004**, *40*, 267–297.

24. Premebida, C.; Nunes, U. Segmentation and geometric primitives extraction from 2D laser range data for mobile robot applications. *Robotica* **2005**, *2005*, 17–25.

25. Xia, Y.; Chun-Xia, Z.; Min, T.Z. Lidar scan-matching for mobile robot localization. *Inf. Technol. J.* **2010**, *9*, 27–33.

26. Ji, J.; Chen, G.; Sun, L. A novel hough transform method for line detection by enhancing accumulator array. *Pattern Recognit. Lett.* **2011**, *32*, 1503–1510.

27. Fernandes, L.A.; Oliveira, M.M. Real-time line detection through an improved hough transform voting scheme. *Pattern Recognit.* **2008**, *41*, 299–314.

28. Nguyen, V.; Martinelli, A.; Tomatis, N.; Siegwart, R. A comparison of line extraction algorithms using 2D laser rangefinder for indoor mobile robotics. In Proceedings of the 2005 IEEE/RSJ International Conference on Intelligent Robots and Systems (IROS 2005), Edmonton, AB, Canada, 2–6 August 2005; pp. 1929–1934.

29. Diosi, A.; Kleeman, L. Uncertainty of line segments extracted from static SICK PLS laser scans, SICK PLS laser. In Proceedings of the Australiasian Conference on Robotics and Automation, Brisbane, Australia, 1–3 December 2003.

30. Soloviev, A. Tight coupling of GPS and INS for urban navigation. *Aerosp. Electron. Syst. IEEE Trans.* **2010**, *46*, 1731–1746.

31. Hancock, J.; Hebert, M.; Thorpe, C. Laser intensity-based obstacle detection. In Proceedings of the 1998 IEEE/RSJ International Conference on Intelligent Robots and Systems, Victoria, BC, USA, 13–17 October 1998; pp. 1541–1546.

32. Iqbal, U.; Okou, A.F.; Noureldin, A. An integrated reduced inertial sensor system—RISS/GPS for land vehicle. In Proceedings of the Position, Location and Navigation Symposium, 2008 IEEE/ION, Monterey, CA, USA, 5–8 May 2008; pp. 1014–1021.

33. Iqbal, U.; Karamat, T.B.; Okou, A.F.; Noureldin, A. Experimental results on an integrated GPS and multisensor system for land vehicle positioning. *Int. J. Navig. Obs.* **2009**, *2009*, 765010.

34. Liu, S.; Atia, M.M.; Karamat, T.; Givigi, S.; Noureldin, A. A dual-rate multi-filter algorithm for LiDAR-aided indoor navigation systems. In Proceedings of the Position, Location and Navigation Symposium-PLANS 2014, 2014 IEEE/ION, Monterey, CA, USA, 5–8 May 2014; pp. 1014–1019.

35. Liu, S.; Atia, M.M.; Karamat, T.B.; Noureldin, A. A LiDAR-aided indoor navigation system for UGVs. *J. Navig.* **2014**, doi:10.1017/S037346331400054X.

36. CHRobotics. UM6 ultra-miniature orientation sensor datasheet. Available online: http://www.chrobotics.com/docs/UM6_datasheet.pdf (accessed on 21 January 2015).

37. SICK. *Operating Instructions: LMS100/111/120 Laser Measurement Systems*; SICK AG: Waldkirch, Germany, 2008.

11

Executed Movement Using EEG Signals through a Naive Bayes Classifier

Juliano Machado [1,2] **and Alexandre Balbinot** [2,*]

[1] Assistive Technology Laboratory, Federal Institute of Rio Grande do Sul (IFSul), General Balbão Street 81, Charqueadas 96745-000, Brazil; E-Mail: julianomachado@charqueadas.ifsul.edu.br

[2] Biomedical Instrumentation Laboratory, Federal University of Rio Grande do Sul (UFRGS), Avenue Osvaldo Aranha 103, Porto Alegre 90035-190, Brazil

* Author to whom correspondence should be addressed; E-Mail: alexandre.balbinot@ufrgs.br

External Editor: Dean M. Aslam

Abstract: Recent years have witnessed a rapid development of brain-computer interface (BCI) technology. An independent BCI is a communication system for controlling a device by human intension, e.g., a computer, a wheelchair or a neuprosthes is, not depending on the brain's normal output pathways of peripheral nerves and muscles, but on detectable signals that represent responsive or intentional brain activities. This paper presents a comparative study of the usage of the linear discriminant analysis (LDA) and the naive Bayes (NB) classifiers on describing both right- and left-hand movement through electroencephalographic signal (EEG) acquisition. For the analysis, we considered the following input features: the energy of the segments of a band pass-filtered signal with the frequency band in sensorimotor rhythms and the components of the spectral energy obtained through the Welch method. We also used the common spatial pattern (CSP) filter, so as to increase the discriminatory activity among movement classes. By using the database generated by this experiment, we obtained hit rates up to 70%. The results are compatible with previous studies.

Keywords: naive Bayes (NB); linear discriminant analysis (LDA); Welch method; brain computer interface (BCI)

1. Introduction

About six decades after the invention of EEG (electroencephalographic signals), studies using brain signals to control devices have emerged, bringing about what we know as BCI (brain-computer interface) or BMI (brain-machine interface). Wolpaw *et al.* [1] point out that brain activity produces electrical signals that can be detected both in invasive and noninvasive ways, and BCI systems can translate these signals into commands, allowing communication with devices without involving peripheral nerves and muscles.

Typically, noninvasive BCI systems use brain activity obtained from the scalp and are capable of allowing basic communication and control for individuals with severe neuromuscular disorders [2]. In general, BCI systems allow individuals to interact with the external environment by consciously controlling their thoughts instead of contracting muscles (e.g., human-machine interfaces controlled or managed by myoelectric signals). They are composed of brain signal acquisition and pre-processing, as well as the extraction of significant features, followed by their classification (see Figure 1). The result of the classification allows external devices to control signals. Another thing about BCI systems is that the user receives stimuli (visual, auditory or tactile) and/or performs mental tasks while the brain signals are captured and processed. Based on the stimulus or task performed by the user, several phenomena or behaviors extracted from the EEG signals can be detected.

Figure 1. A typical system block diagram.

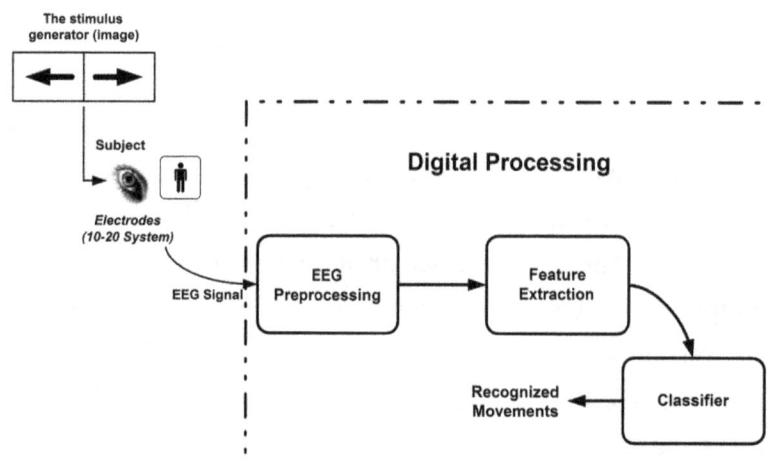

In practice, physiologically meaningful EEG features can be extracted from several frequency bands of recorded EEG signals. Therefore, many electrical brain activities have been used in EEG-based BCI systems, e.g., μ rhythm [3–7], slow cortical potential [8], event-related P300 [9,10] and steady-state visual evoked potential [11,12]. The activity most widely used to monitor the brain for BCI applications is the μ rhythm, which is related to motor actions [2,3,13,14]. Unlike event-related brain activities, the μ rhythm can be voluntarily modulated by users.

Defined as the mental simulation of a kinesthetic movement [15,16], the imaginary motor activity can also modulate μ rhythm activities in the sensorimotor cortex without any physical body movement. McFarland *et al.* [17] reported that imagined movement signals can be reflected in the β rhythm (13–22 Hz). Pfurtscheller [18] pointed out that both α (8–13 Hz) and β rhythm amplitudes might serve as effective input for the BCI system to recognize patterns of real or imaginary movement. The event-related desynchronization (ERD), which is the reduction of a specific frequency energy component, refers to increased neural activity during performed or imagined movements and appears over the primary motor cortex; the event-related synchronization (ERS), which is the enhancement of a specific frequency energy component, refers to a neural suppression during non-performed or non-imagined movements and sometimes appears over the primary motor cortex [14,18–21]. Following an ERD that occurs shortly before and during the movement, an enhancement of β oscillations (β ERS) appears within a one-second interval after the movement offset [20]. Such a post-movement β ERS has been witnessed after voluntary hand movements [22–24], passive movements [25], imagined movements [26] and movements induced by functional electrical stimulation.

The study of Müller *et al.* [26] compares ERD/ERS patterns during active and passive foot movements, both in healthy individuals and paraplegic patients suffering from a complete spinal cord injury. The results showed mid-central β ERD/ERS patterns during active, passive and imagined foot movements in healthy individuals against a diffuse and broad distributed ERD/ERS pattern. During active foot movements in paraplegic patients, only a single patient showed similar ERD/ERS patterns related to active movement, and no significant ERD/ERS patterns were observed in paraplegic patients during passive foot movement.

Similar results can be found related to both real and imaginary hand movement [27]; also, it is known that the brain area that controls hands has a good spatial separation [20]. Accordingly, it is possible to distinguish certain hand movements while processing the EEG signal's electrical parameters. Based on this hypothesis, we aim to study and develop a system that uses EEG signals acquired from surface electrodes to describe the movement of the human hand. This study examined the behavior of spectral estimation by using a periodogram and the signal's energy, so as to verify whether the extracted features can be used both in a naive Bayes (NB) and linear discriminant analysis (LDA) classifier. Furthermore, NB and LDA classifiers were compared for both a recognized international database [28] and the data collected in an uncontrolled environment.

2. Methods

2.1. Pre Processing

2.1.1. Spectral Estimation

Due to signal modulation during movement, it is established practice to use the signal on the frequency domain as a feature for the movements' classification [29]. A classic estimator for the spectral energy would be the Fourier transform (FT) of the signal's autocorrelation function [30]; however, estimators of the signal frequency spectrum should consider the signal's non-stationary behavior. Thus, it is necessary to apply a method considering its nonstationarity, such as the wavelet transform, or to take an isolated segment and consider it as weakly stationary [31].

In this study, all EEG signals were windowed so that loss of resolution in the frequency domain and the influence of the lateral lobes of the window are the main consequences of applying a window to the signal [32]. The resolution is directly influenced by the main lobe; the leakage adds a tendency to the estimator in the frequencies adjacent to the frequency of interest [33]. According to [32], the approximate width of the central lobe in the rectangular window varies according to its size, getting narrower as more samples are added to the window. By using an ANOVA (analysis of variance) table, we have also observed how the size of the window can affect the classification.

The fact that the EEG signal can only be considered stationary in short periods of time (from about 1 to 2 s) can become an issue for the experiment. One solution is to use Welch's modified periodogram [33,34], which overlaps the segmented windows up to 50% of their size, allowing a large number of windows. The Welch's method (also known as the periodogram method) for estimating power spectral density is carried out by dividing the time signal into successive blocks, forming the periodogram for each block and averaging them. Equation (1) defines the m-th windowed, zero-padded frame from the signal x:

$$x_m(n) = w(n)x(n + mR); n = 0, 1, \dots, M - 1; m = 0, 1, \dots, K - 1 \quad (1)$$

where R is defined as the window hop size, M is the sample size, and K denotes the number of available frames. Then, the periodogram of the m-th blocks is given by:

$$P_{x_m,M}(w_k) = \frac{1}{M}\left|FFT_{N,k}(x_m)\right|^2 = \frac{1}{M}\left|\sum_{n=0}^{N-1} x_m(n)e^{-\frac{j2\pi nk}{N}}\right|^2 \quad (2)$$

where FFT is the Fast Fourier Tranform. In other words, the Welch estimator of the power spectral density is given by:

$$S_x^W(w_k) = \frac{1}{K}\sum_{m=0}^{K-1} P_{x_m,M}(w_k) \quad (3)$$

With a smaller number of windows, it is possible to obtain a lower random error, as given by Equation (3), since it increases the number of degrees of freedom. According to Stoica and Moses [33] empirical results showed that this method reduces the variance.

2.1.2. Spatial Filter

One of the challenges of using classification algorithms is the large amount of generated features against the amount of available training data. In the field of machine learning, we call this the curse of dimensionality. Therefore, the EEG signal pre-processing aims to reduce the space of features by selecting only the most discriminatory ones from the states to be classified [31].

Performed or imagined movements create certain spatial patterns in the scalp, generating in the central cortex a desynchronization contralateral to the movement followed by an increase of energy ipsilateral to the movement. Thus, it is possible to select channels that best discriminate right and left hand movements instead of using all channels, which reduces the space of features [35]. We can define the energy of a band pass-filtered signal by its variance [30], which, therefore, becomes perfect for capturing the discriminatory effect of the EEG signal within both α and β rhythms during the ERD and ERS related to the voluntary hand movement.

We have suggested the usage of the common spatial pattern (CSP) algorithm in order to maximize the discriminatory activity between two classes of EEG signals. As the scalp conducts elements not belonging to the EEG signal, such as myoelectric signals from face muscle activity, the channels show a lot of similar nondiscriminatory activity covering up the EEG discriminatory activity. This happens mainly because the cortex signal is weak (amplitude in the μV band) in relation to the myoelectric signals (amplitude in the mV band). The maximization of the discriminatory activity can be accomplished through a linear transformation that maximizes the variance of one condition, while minimizing the variance of the other condition by moving the original sensor space to a new one. In this paper, it is important to notice that bold uppercases represent a matrix, while bold lowercases represent a one-dimensional vector. For our purposes, an EEG signal segment is considered as a band pass-filtered signal, of size T with C channels, represented by $X \in \mathbb{R}^{C \times T}$ (or $x(t) \in \mathbb{R}^C$) in a certain time t. Thus, X is a concatenation of signals $x(t)$ represented by:

$$X = [x(t), x(t+1), \ldots, x(t+T-1)] \tag{4}$$

To be more precise, let $X_{(k)}^n$ be a point of the bandpass-filtered EEG signal segment with size N of the class (k), defining the estimator of the co-variance matrix:

$$\Sigma^{(k)} = \frac{1}{N} \sum_{n=1}^{N} \frac{X_{(k)}^n \times X_{(k)}^n{}'}{\text{trace}(X_{(k)}^n \times X_{(k)}^n{}')} \tag{5}$$

Considering two classes ($\Sigma^{(+)}$ and $\Sigma^{(-)}$), the CSP analysis consists of calculating a matrix W and a diagonal matrix $\Lambda^{(k)}$ with elements in $[0,1]$, such that:

$$W^T \times \Sigma^{(+)} \times W = \Lambda^{(+)}$$
$$W^T \times \Sigma^{(-)} \times W = \Lambda^{(-)} \tag{6}$$
$$(\Lambda^{(+)} + \Lambda^{(-)} = I)$$

where I is the identity matrix.

To accomplish this, it is necessary to whiten the matrix $\Sigma = \Sigma^{(+)} + \Sigma^{(-)}$ as follows:

$$P\Sigma P^T = I \tag{7}$$

This decomposition is always possible due to the positive definiteness of Σ. Next, we shall transform the covariance matrices of each class, $S^{(+)} = P \times \Sigma^{(+)} \times P^T$ and $S^{(-)} = P \times \Sigma^{(-)} \times P^T$ and find an orthogonal matrix U and a diagonal matrix $\Lambda^{(k)}$ by the spectral theory, such that:

$$S^{(+)} = U \times \Lambda^{(+)} \times U^T \text{ and } S^{(-)} = U \times \Lambda^{(-)} \times U^T \tag{8}$$

The spatial filter W projecting the signal $x(t)$ from the original sensor space to the surrogate space $x_{CSP}(t)$ is then given by the projection of matrix P by U':

$$W = (U' \times P)' \tag{9}$$

The new sensor space is generated through a supervised decomposition of the signal $x(t)$ parameterized by a matrix $W \in \mathbb{R}^{C \times C}$ that projects the signal to a surrogate space $x_{CSP}(t) \in \mathbb{R}^C$:

$$x_{CSP}(t) = W^T \times x(t) \tag{10}$$

Therefore, notice that each column $w_j \in \mathbb{R}^C$ ($j = 1, \ldots, C$) consists of a spatial filter that linearly recombines all channels' components, creating a new channel. Furthermore, notice that $A = (W^{-1})^T$ is the matrix leading to the original sensor space once again by giving the spatial pattern of the signal $x(t)$ [36]. The columns w_j resulting in the values of $\Lambda_j^{(c)}$ closest to 1 in both classes will be those that best discriminate them.

2.2. Signal Classification

2.2.1. Naive Bayes

A classifier always aims to reach the best hypothesis H through a given training dataset. The Bayes theorem allows one to calculate the *a posteriori* probability (the probability of a hypothesis considering a variable's value) based on the *a priori* probability (the frequency of each hypothesis) of both the data found and the total data, according to Equation (11) [37]:

$$P(v_j|A) = \frac{P(A|v_j) \times P(v_j)}{P(A)} \tag{11}$$

where v_j is the hypothesis j in the set of hypotheses V, and A is the set of attributes $< a_1, a_2, ..., a_n >$ describing the data.

When A has more than one attribute, it is then necessary to estimate $P(a_1, a_2, ..., a_n|v_j)$ in order to calculate $P(v_j | a_1, a_2, ..., a_n)$. The problem is that to estimate $P(A | v_j)$, it is necessary to have an extremely large amount of samples. Moreover, it is computationally costly, since it is necessary to calculate the joint probabilities for all possible A [30,37]. Thinking about that, we suggested the use of the NB classifier, which assumes that all attributes in A are independent. There is literature discussing that even if these attributes are not totally independent, it is possible to obtain a good classification performance. In addition, it has a simple implementation [38,39]. Thus, the joint probability is given by:

$$P(a_1, a_2, ..., a_n|v_j) = \prod_i P(a_i|v_j) \tag{12}$$

and the classifier output is given by:

$$v_{MAP} = \operatorname*{argmax}_{v_j \in V} \left\{ P(v_j) \times \prod_i P(a_i|v_j) \right\} \tag{13}$$

where v_{MAP} is the maximum *aposteriori* probability calculated within the space of hypotheses V. Notice that it is only necessary to estimate the probability distribution of each attribute for each class, it not being necessary to calculate $P(A)$ if the number of observations is the same for each class.

2.2.2. Linear Discriminant Analysis

The LDA aims to project the data into a hyperplane within the space of features to find the orientation resulting in the projection that best discriminates both classes [38,39]. A linear discrimination combining the components of the features space $x \in \mathbb{R}^D$ can be cast as:

$$g(x) = p^T \times x + p_0 \tag{14}$$

where $p \in \mathbb{R}^D$ is the weight vector, p_0 is a constant and D is the size of the feature vector. A linear classifier for the classes v_1 and v_2 establishes the following decision rule:

$$\text{chooses } v_1 \text{ if } g(x) > 0$$
$$\text{chooses } v_2 \text{ if } g(x) < 0 \tag{15}$$

Accordingly, the class v_1 is chosen when the internal product is superior to $-w_0$, while the class v_2 is chosen when it is inferior. The hyperplane is given by the normal vector $\boldsymbol{p}$ and its orientation is given by the vector $\boldsymbol{p}$ that maximizes the function J:

$$J(\boldsymbol{p}) = \frac{\boldsymbol{p}^T S_d \boldsymbol{p}}{\boldsymbol{p}^T S_c \boldsymbol{p}} \tag{16}$$

where S_d and S_c are the data co-variance matrices of one class, so that S_c is the common co-variance of all classes. We calculated $\boldsymbol{p}$ in order to maximize the function J and find a vector $\boldsymbol{p}$ that maximizes the discriminatory activity between classes regarding the common activity [38,39].

3. Experimental Section

3.1. Materials and Data Synchronization

The proposed experimental BCI system is shown in Figure 2. We have based our experiment on computer-generated stimuli introduced by Monitor 2 to a volunteer who remained sitting in a chair. Data were first obtained from a 10–20 system EEG cap (Spes Medica, CAMSUMA20, Genova, Italy) [40] and then analogically amplified and filtered by an EEG to be digitally converted for the computer through an ADC (Analog to Digital Converter) from National Instruments (NI USB 6008, National Instruments, Austin, TX, USA). A key was placed both on the right and left arms of the chair to be triggered according to the stimulus; the signal reading was held through another ADC (NI USB 6008).

Both stimuli generation and data acquisition were synchronously performed by the computer shown in Figure 2. The acquisition was continuously performed until the ensembles were divided into 8-s windows. Thus, the stimuli presentation was synchronized in time to the signal by software. The stimuli sequence shown in Figure 3 was submitted as the following:

(1) From 0 to 1.5 s: a white screen appeared establishing the so-called reference period;
(2) From 1.5 to 3 s: a pre-stimulus a cross appeared on the screen;
(3) From 3 to 6 s: the stimulus presentation occurred (a blue arrow pointing to the right or a red arrow pointing to the left);
(4) From 6 to 8 s: a white screen appeared once again establishing the so-called post-stimulus period.

This experiment was based on previous studies [22,28] that proved that there exists a large bilateral desynchronization within the frequency bands of both μ and β rhythms during movement imagination, which always keep the energy ipsilateral to the movement superior to the energy contralateral to the movement. Such studies proved that during brief periods (from 1 to 3 s), the movement imagination shows a difference in energy within μ and β rhythms that is capable of distinguishing a movement from another. The choice of using the 8-swindow occurred experimentally.

Along the experiment, the volunteer was asked to push the mechanical keys installed on the chair. Every time a red arrow appeared on the screen, indicating the left-hand movement, the volunteer had to push the key located on the left arm of the chair using his left hand. Whenever a blue arrow appeared, he/she had to perform the same action, but this time using his/her right hand. Such an experiment was held in a synchronous way, *i.e.*, it was possible to control the time when the movement was performed [41], supporting the identification of the ERS/ERD effect during the signal analysis.

A post-stimulus period was generated in order to allow the brain enough time to return to its normal state after movement performance [42].

Figure 2. Block diagram of the proposed experiment.

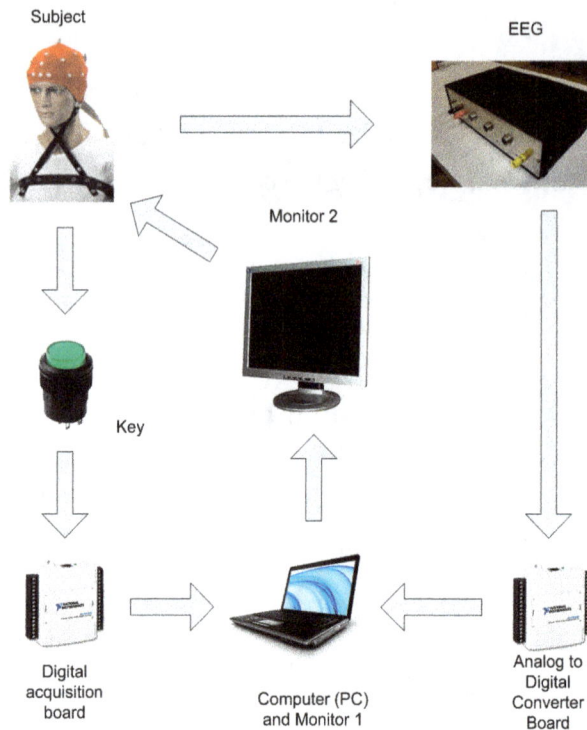

Figure 3. Ensemble time scale.

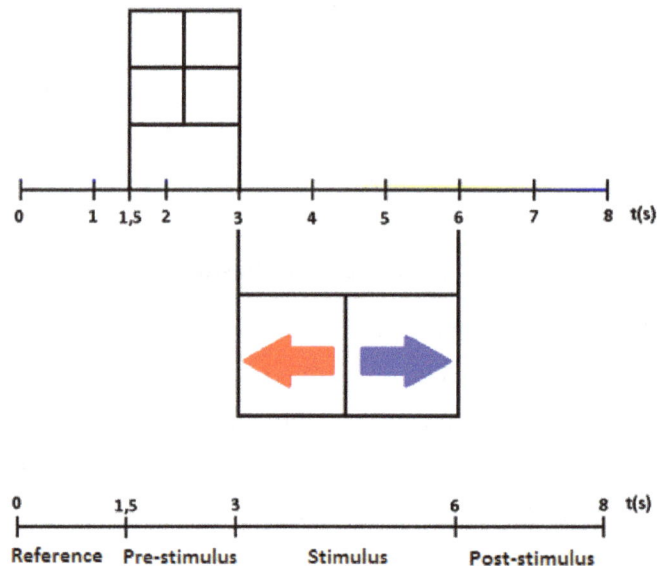

The ADC (NI USB6008) was configured to acquire 6 EEG channels (F3, F4, C3, C4, P3 and P4) with a sample frequency of 256 Hz. These channels were selected to cover the most important areas related to the motor cortex. For data configuration, acquisition and synchronization, we developed two software systems through the LabVIEW (Version 2009) development tool. One works like a control software (Software I) and is responsible for calling out the other (Software II), as well as gathering data and managing the digital-analog acquisition (Figure 2). The acquisition must be synchronized with stimuli presentation performed by Software II (Monitor 2), which is responsible for triggering the acquisition process by communicating with Software I (Monitor 1). Next, the presentation of a pre-determined number of ensembles starts as Software I saves the corresponding data; the sequence generating stimuli is randomly created and differs from one experimental run to another. Moreover, Software II also controls and saves data obtained from the keys. For this experiment, key control was performed, so that the volunteer could have a 1-s window after stimulus presentation (from 3 to 4 s) to push the key, so as to guarantee that it did not react if pushed before or after the appearing of the window.

The synchronization between stimulus presentation and time base (where the first sample equals 0 s and the n sample equals $n \times T$ s; T s being the sample period) was tested through feedback in the acquisition board. The acquisition board has analog inputs for A/D conversion, as well as digital inputs and outputs at TTL (Transistor-Transistor Logic) voltage levels. The feedback consisted of linking 2 digital outputs (S_0 and S_1) to 2 analog channels and codifying each segment through digital outputs. A small change was implemented in the Software II, so as to cause the digital output values to change according to a certain established pattern when changing the ensemble segment, e.g., as in the passage from the reference period to pre-stimulus.

The outputs connected to the 2 analog channels remained in their logic states during the whole period of the observed segment. Along the stimulus period, for instance, S_0 remained at a high logic level (5 V), while S_1 remained at a low logic level (0 V) during all 3 s. Therefore, it was possible to verify whether the transitions and the time base of data acquisition occurred synchronously by checking the data obtained through analog channels.

3.2. EEG Signal Processing

EEG signal preprocessing consists of applying a band pass filter both in μ and β rhythms (4th order Butterworth digital filter) and separating signals into ensembles according to each movement class (right or left). The Software I registers both the synchronism data and the pushed button. Data are then separated regarding both the class and the pushed button, e.g., in the case of the synchronism file indicating an ensemble pointing the arrow to the left and the volunteer pushes the right button or no button at all, the segment acquired is then automatically discarded.

CSP Filter

For our purposes, the first step to calculate the spatial filter W is to estimate the co-variance matrices of the training set using Equation (6). It is important to notice that every time the CSP filter was calculated, we used only the training set.

In order to maximize the discriminatory activity between the two classes during hand movement, we applied both 1- and 2-s windows to the EEG signal to further extract the channels' spectral components and variances that best discriminate the classes in terms of the determined eigenvalues. According to previous studies, the ERD/ERS effects usually occur up to 2 s after movement performance [27,41,42]. Thus, we considered times 3 and 4 s of the ensemble shown in Figure 3 as the starting points of the window. Accordingly, to calculate W we need to:

(1) Estimate $\boldsymbol{\Sigma}^\mathbf{L}$ and $\boldsymbol{\Sigma}^\mathbf{R}$ (which are the co-variance matrices for the left and right classes, respectively) through the training set;

(2) Find the matrix $\boldsymbol{\Sigma} = \boldsymbol{\Sigma}^\mathbf{R} + \boldsymbol{\Sigma}^\mathbf{L}$;

(3) Perform the "whitening" operation in order to obtain the matrix $\boldsymbol{P}$;

(4) Decompose $\boldsymbol{\Sigma}^\mathbf{L}$ and $\boldsymbol{\Sigma}^\mathbf{R}$ through matrix $\boldsymbol{P}$ to obtain the matrices $\boldsymbol{S}^\mathbf{L}$ and $\boldsymbol{S}^\mathbf{R}$, whose eigenvalues $\boldsymbol{\Lambda}^\mathbf{L}$ and $\boldsymbol{\Lambda}^\mathbf{R}$ represent the discriminatory activity in the new CSP channel space;

(5) Select both the n largest eigenvalues $\boldsymbol{\Lambda}^\mathbf{L}$ and $\boldsymbol{\Lambda}^\mathbf{R}$ that will maximize the variance in the left-hand movement condition while minimizing the variance in the right-hand movement condition;

(6) Calculate the spatial filter and select $2 \times n$ columns of the matrix $\boldsymbol{W}$, which are related to the n largest eigenvalues of $\boldsymbol{\Lambda}^\mathbf{L}$ and $\boldsymbol{\Lambda}^\mathbf{R}$, respectively.

To successfully accomplish the EEG signal classification, we need to properly choose the features; that is, for the LDA classifier, the energy of the two best CSP channels (for the NB classifier, Welch periodogram's components are also used) that allow identifying and classifying both right- and left-hand movements. Many authors use algorithms to verify each feature's relevance [43]. For this experiment, the manual method for choosing features performed by an expert was applied. Using an expert to virtually select the best features is a common practice that many times generates higher hit rates than automatic methods [41,43].

3.3. Features Extraction

It is important to notice that all of the features were selected using only the training set.

3.3.1. Energy of CSP Filtered EEG

The feature extracted through CSP filters is the logarithm of the energy of the signal projected with the best eigenvalues for each class. Accordingly:

$$feature_j = \log\left(w_j^T \times X \times X' \times w_j\right) \tag{17}$$

where $feature_j$ is the feature vector used for classification and w_j is the j-th spatial filter. Operation $\log()$ is applied in order to approach energy distributions as close as possible to a Gaussian. The EEG signal segment X is obtained considering previous discussions. The expert chooses the features for the column of the filter W with the highest discriminatory activity.

3.3.2. Welch's Periodogram Components

The first step to determine the spectral components was to find the initial instant of the cut of the window, by using the training set. It is well known that the ERD/ERS effects can occur during movement and/or movement planning, mainly in μ and β rhythms, where the signal is filtered. After band pass filtering it, the signal was squared to obtain its energy. Next, we calculated the energy average of all ensembles in each class to verify the ERD/ERS evidence, as described by [27,42]. The average energy of each channel is given by:

$$Energy = \frac{1}{M}\sum_{i=0}^{M} C_{ji}^{(c)^2} \tag{18}$$

where $C_j^{(c)}$ is the channel j of the EEG signal in condition (c) and M is the number of ensembles. To better visualize the data, a moving average filter was applied to the resulting signal, so as to smooth it. Results show the time instant in which the ERD/ERS occurs, determining the initial instant to be windowed. Effects are expected to be more prominent from the instant of the movement to about 2 s after it. Once the initial instant of the "windowing" was obtained, Equation (3) was applied within a 2-s window. For a 2-s window at 50% overlap, we used three 1-s windows, as shown in Figure 4.

Figure 4. Windowing for a 2-s segment at 50% overlap between windows according to the Welch method.

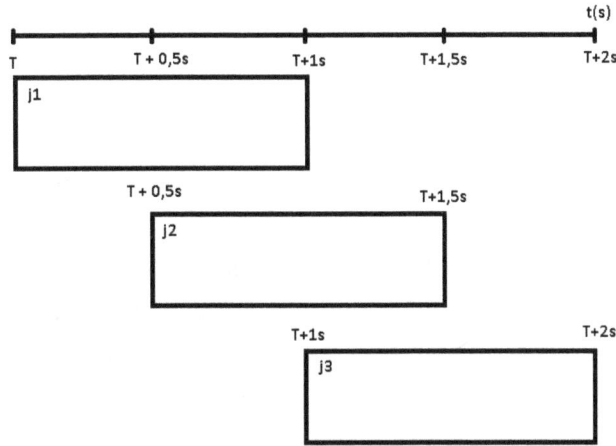

By using 1-s windows, it was possible to obtain a frequency resolution of 1 Hz between each spectral component. Values were represented by graphs, as well as the lateralization index (*LI*), which shows the difference of energy between the right and left hemispheres [27,44]:

$$LI = \left[\left(EnergyLCH_{leftmovement} - EnergyRCH_{leftmovement}\right) + \left(EnergyRCH_{rightmovement} - EnergyLCH_{rightmovement}\right)\right]\Big/2 \quad (19)$$

where the *Energy LCH* of the channels in the left hemisphere and *Energy RCH* is the energy of channels in the right hemisphere. *Energy LI* values indicate a high contralaterality of the signal. By applying them to the energy of each frequency component, it was possible to observe which components best discriminate the classes. The *LI* can also be obtained in the time domain, so as to verify in which instants the highest *LI* values occur, indicating a larger discriminatory activity. The expert uses the components with high *LI* to compose the set of features.

Moreover, same spectral components used by [29,45–47] to classify EEG signals are suggested as features. In order to reach a good spectral component resolution, a rectangular window was used, since it possesses the lower central lobe. We separated it into three 1-s windows at 50% overlap between each window, as shown in Figure 4. The feature vector is given by:

$$\boldsymbol{feature_F} = \log(\widehat{Gxx}(F)) \quad (20)$$

where $\boldsymbol{feature_F}$ is the feature vector used for signal classification and $\widehat{Gxx}(F)$ is the vector containing the estimated frequency components. Notice that it is not necessary to use all frequency components, but only the best ones. It is important to verify that operation log() was applied in order to normalize the distributions. Time T (from the beginning of the "windowing") was determined according to the considerations mentioned before.

In this study, both NB and LDA classifiers were used. The NB classifier is responsible for modeling the distributions. Since features were normalized through the log() function, the normal distribution was used. The features' vectors and quantity provide the number of distributions. For comparison, the LDA classifier was used in order to determine both the vector p_j and the constant p_0. The number of constants of the vector p_j is given by the amount of features used. It is important to point out that using large amounts of features in a linear classifier can cause an overfitting [37].

4. Results and Discussion

Our findings came from ten volunteer sin an uncontrolled environment.

4.1. Analysis Based on Signal Energy

Through the signal energy and the *LI* analysis over time, it was possible to identify when the ERD/ERS occurred and to determine when the signal should be windowed in order to extract its features, so as to better discriminate movement classes. An analysis of the *LI* average energy of all of the EEG signal ensembles over time was held. The experiment had four sessions performed (S_1 to S_4) of up to 140 ensembles. However, the amount of ensembles varied according to the movement, since only the ensembles in which the volunteers had pushed the correct button were selected.

As an example, Figures 5 and 6 show a graphic result of the analysis of the relative average energy of the signal for C3 and P3 channels and C4 and P4 channels, respectively, in session S_3. Both channels were band pass filtered in the µ band (from 8 to 12 Hz), and a 63-point moving average filter was applied over the average signal, so as to smooth the graph lines. In the *y*-axis is shown the relative energy between the reference times, while the *x*-axis shows time periods. Accordingly, it was possible to visualize how much the energy increases or decreases between no activity at all (reference period) and the performed movement (stimulus).

Figure 5. Relative average energy of channels C3 and P3 during S_3 (µ rhythm).

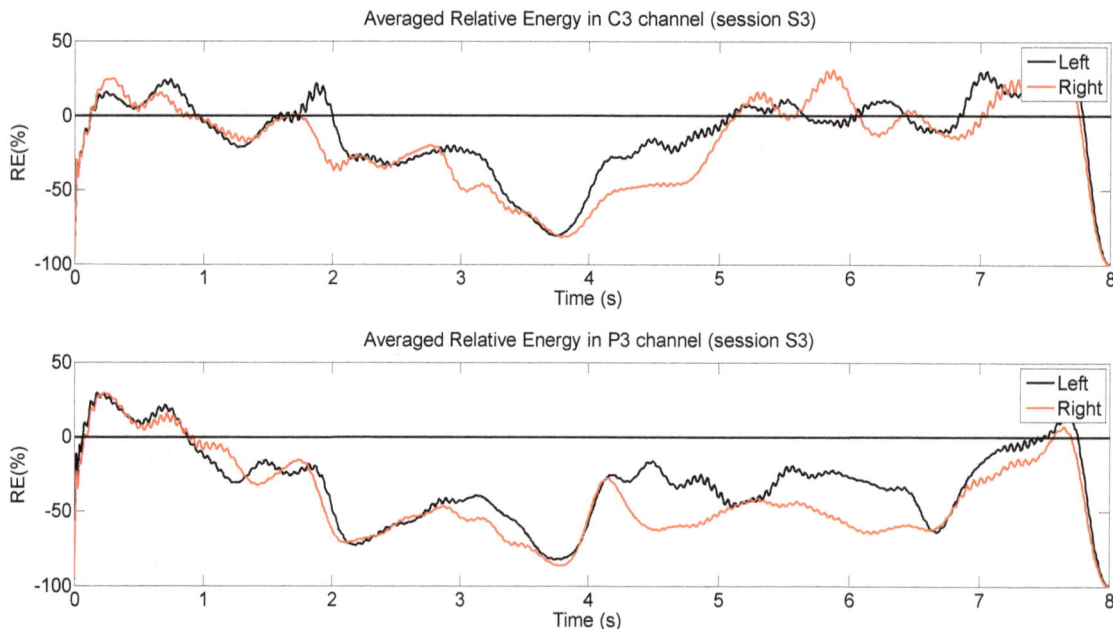

Figure 6. Relative average energy of channels C4 and P4 during S_3 (μ rhythm).

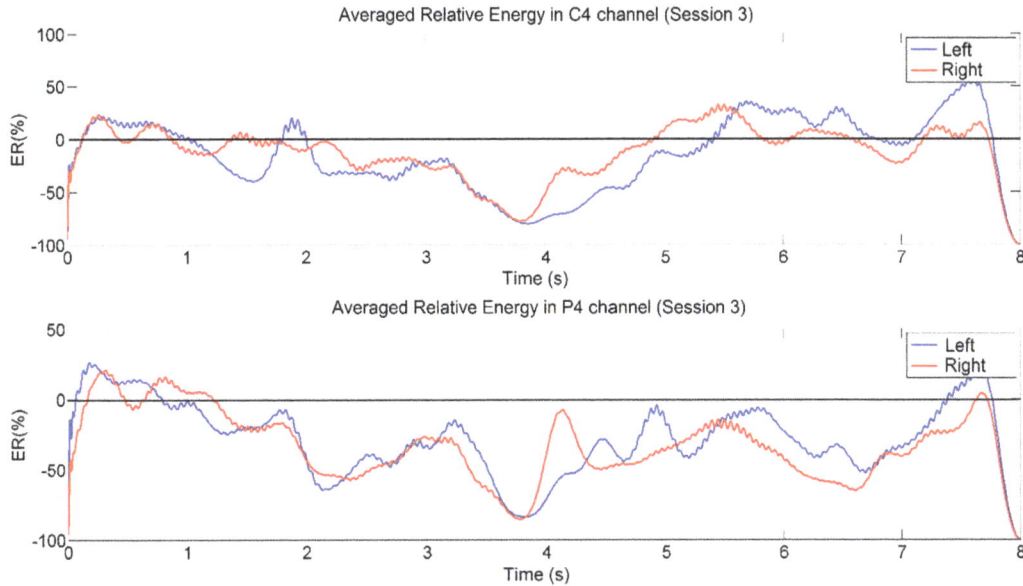

Results show a strong desynchronization about 500 ms after pre-stimulus presentation at 1.5 s in channels C3 and P3 for both types of movement (Figure 5). ERD/ERS effects can be observed both during and after movement in the channel covering the motor cortex (C3), always keeping the energy ipsilateral to the movement superior to the energy contralateral to the movement; however, an ERS appeared to be sharpest in the left hemisphere. The same effect had been observed in channels covering the parietal lobe, indicating motor activity. Notice that in all cases, the ERS effect of the ipsilateral side occurred about 800 ms after the arrow pointed at time 3 s, indicating a discriminatory effect between the energy of both classes. Thus, in order to extract the spectral components and the signal energy for signal windowing, we used initial values between 3.5 and 4 s (from 500 ms to 1 s after the appearance of the arrow). No discriminatory activity was observed in β rhythm and channels F3 and F4. Therefore, no signal from β rhythm was used for classification process, since it showed no relevant discriminatory feature.

4.2. Analysis Based on LI

The *LI* was calculated through central channels (C3 and C4) and parietal channels (P3 and P4) only, using Equation (20), since they showed motor activity. Figure 7 shows the normalized *LI* average graph over time for all four sessions. *LI* results indicate a high signal lateralization on the central lobe from about 500 ms after the appearing of the arrow to about 2 s after its appearing at 5 s, corroborating the choice of starting the windowing at 3.5 s. The graph is normalized, and also note that when the curves are above the zero line, this indicates that an ERD is occurring; when below zero, this indicates that an ERD is occurring, and when is below zero, this indicates that an ERS is occurring, according to Equation (20).

4.3. Analysis Based on the Welch Periodogram

This session shows the results obtained from the periodogram. The *LI* (see Figure 8) of the spectral components was also evaluated to determine which frequency components presented a larger discrimination. A 2-s window was used to obtain a frequency resolution of at least 1 Hz; signals were filtered by a band pass between 8 and 24 Hz, so as to entirely cover the sensorimotor rhythm. Only as an example, Figure 9 shows the analysis in the frequency of the experiment in session S_3. Our findings show a clear energy distinction between the movement classes in μ rhythm on channel C3 and a slightly distinction on channel P3, always keeping the energy ipsilateral to the movement superior to the energy contralateral to the movement. We found similar results for channels C4 and P4. No relevant discriminatory activity was observed within β rhythm, since, as mentioned before, there is no perceptive discriminatory activity in β rhythm regarding the volunteers.

Figure 7. Relative average *LI* for channels C3, C4, P3 and P4 in all sessions.

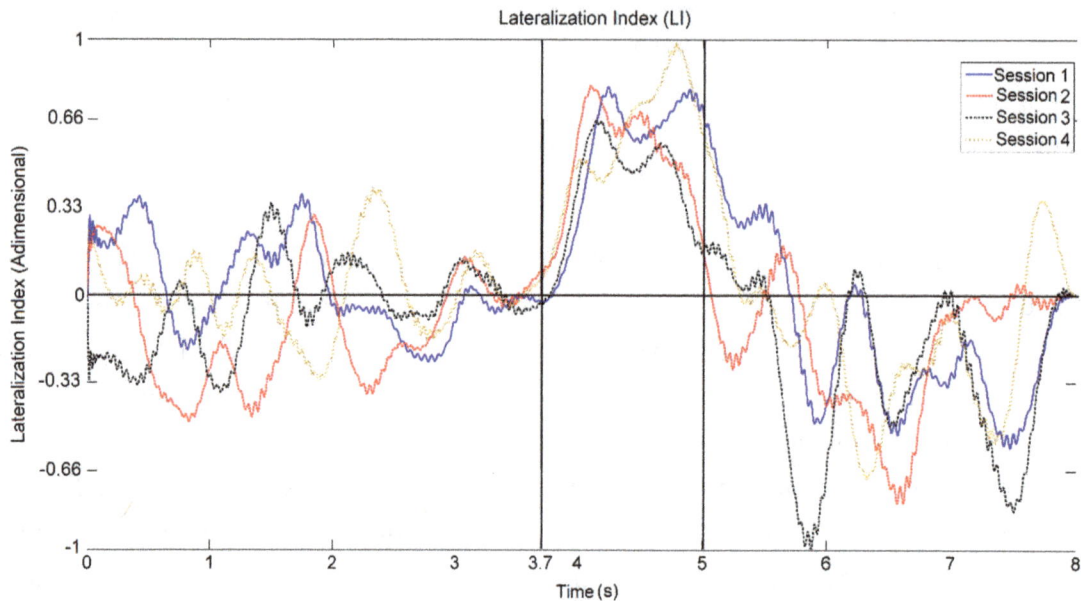

Figure 8. *LI* on the frequency domain for all sessions.

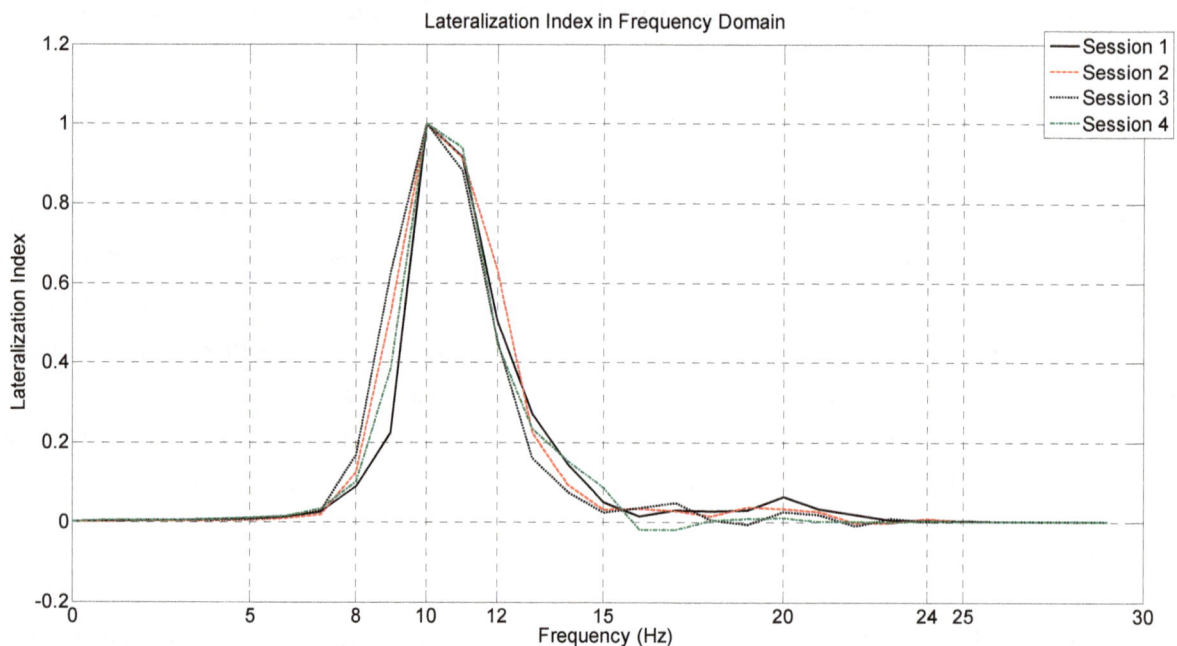

Figure 9. Average energy on the frequency domain for channels C3 and P3.

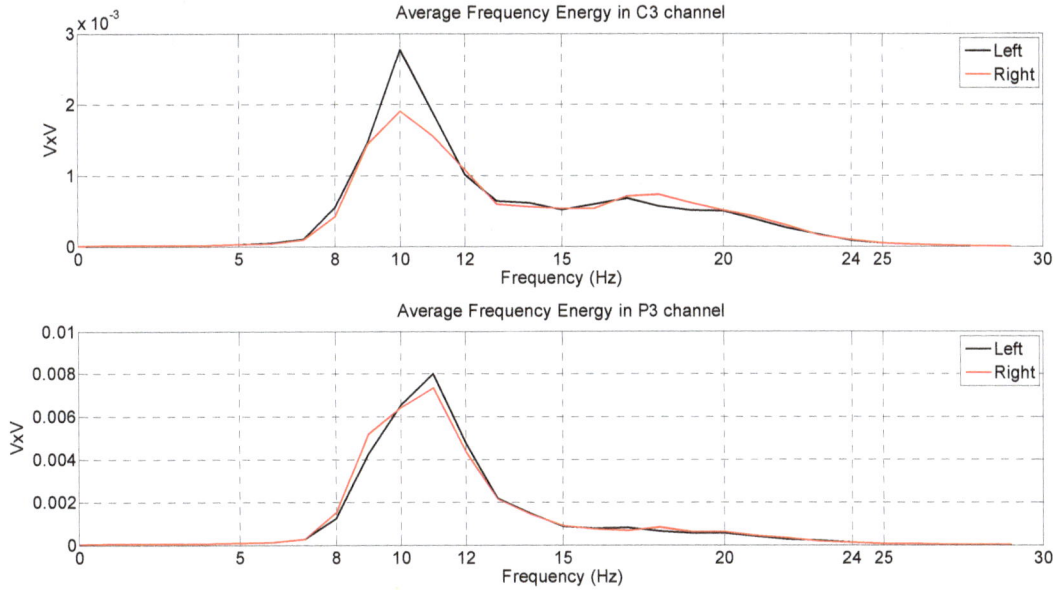

4.4. Analysis Based on the Spatial Filter

In order to analyze the CSP filter, its coefficients had to be calculated based on the entire database and then applied to it. Filters were calculated through a two-second window starting at 3 s. These windows were chosen, because an evaluation of the filter functionality showed whether the segment of the signal creates a new channel space with discriminatory activities by analyzing the selected segment. As mentioned before, only central and parietal channels were considered for calculating the coefficients of the filter. Signals were filtered within μ rhythm. Original channels (C3, C4, P3 and P4) were linearly combined by the filter W obtained according to our methodology, so as to create a new space of channels, which were named CSP1, CSP2, CSP3 and CSP4. As an example, Figure 10 shows the average energy in the frequency domain for channels CSP1 and CSP2.

Figure 10. Average energy in the frequency domain for channels CSP1 and CSP2. CPS, common spatial pattern.

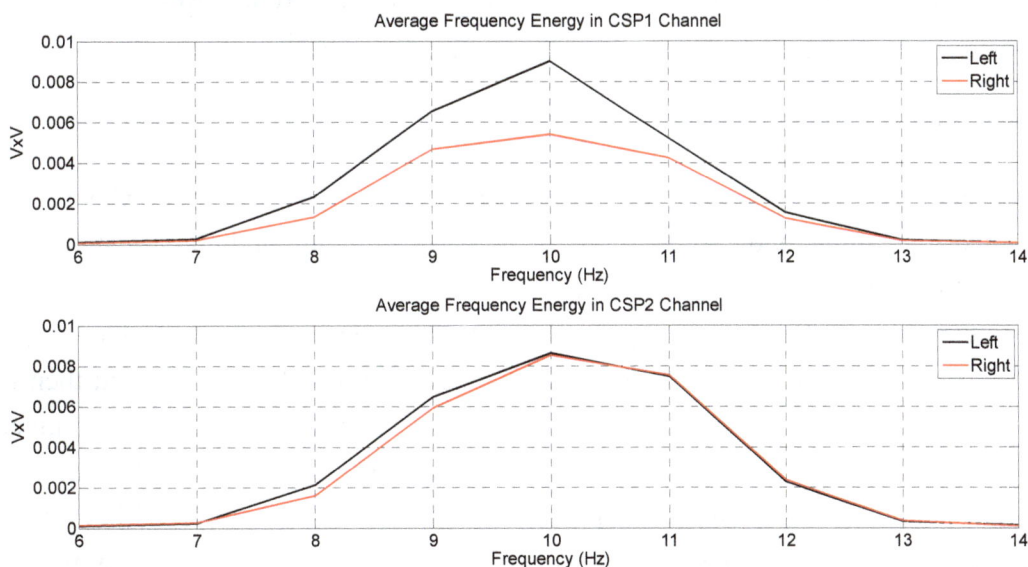

The eigenvalues matrices $\mathbf{\Lambda}^{(R)}$ and $\mathbf{\Lambda}^{(L)}$ indicate which channels present the highest discriminatory activity. Notice that the results obtained from $\mathbf{\Lambda}^{(R)}$ and $\mathbf{\Lambda}^{(L)}$ are in agreement with the graphic results. In channel CSP1 ($\mathbf{\Lambda}^{(k)}$ first column), $\mathbf{\Lambda}^{(R)}$ has minimum value and $\mathbf{\Lambda}^{(L)}$ has maximum value, indicating that the left movement class energy should be superior to the right movement class energy (also, $\mathbf{\Lambda}^{(R)} + \mathbf{\Lambda}^{(L)} = I$):

$$\mathbf{\Lambda}^{(R)} = \begin{matrix} 0.37 & 0 & 0 & 0 \\ 0 & 0.47 & 0 & 0 \\ 0 & 0 & 0.50 & 0 \\ 0 & 0 & 0 & 0.54 \end{matrix}$$

$$\mathbf{\Lambda}^{(L)} = \begin{matrix} 0.63 & 0 & 0 & 0 \\ 0 & 0.53 & 0 & 0 \\ 0 & 0 & 0.50 & 0 \\ 0 & 0 & 0 & 0.46 \end{matrix}$$

$$(21)$$

The first and only column of matrices $\mathbf{\Lambda}^{(R)}$ and $\mathbf{\Lambda}^{(L)}$ shows the filters that best discriminate both classes, which is also in agreement with our graphic results. Applying the filter had also increased the discriminatory activity between both movement classes. We could verify this by analyzing the proportional energy between both classes for the same channel, according to:

$$E_P^{(c)} = \frac{E^{(c)}}{E^{(R)} + E^{(L)}} \tag{22}$$

where $E_P^{(c)}$ is the proportional energy in class (c), $E^{(c)}$ is the energy of class (c) and $E^{(R)}$ and $E^{(L)}$ are the energies of the right class movement and the left class movement, respectively. By verifying the proportional energy in the right class for the 10-Hz component for the channel C4 of the original sensor space, for instance, we obtained the results shown in Table 1.

Thus, filter application increased the contrast between both classes according to the results. It is important to point out that the coefficients were estimated from all databases; for classification, however, the training set only was used.

Table 1. A comparison of the proportional energy between classes with and without the CSP filter.

Without CSP Filter, Channel C4	With CSP Filter, Channel CSP4
$E_P^{(R)} = 0.59$	$E_P^{(R)} = 0.63$

4.5. Classification Based on Signal Energy

Results shown in this session were obtained from the cross-correlation procedure, where 9/10 of the data were separated for training and 1/10 for classification. The training and classification procedure were performed 100 times; both training and testing data were selected randomly. The classification rate was given by the average of all performances. The standard deviation is also presented.

The spatial filter was calculated first through four channels (C3, C4, P3 and P4) and then through two channels (C3 and C4). Calculating the filter through four channels allowed a new channel space with four CSP channels, where the two columns of filter W with the highest eigenvalues of each class were tested. All resulting channels (CSP channels) were considered. Calculating the filter through two channels allowed two new CSP channels to be used to calculate the classifier. As an example, Table 2 shows all four resulting CSP filters. Following the same methodology, Table 3 shows the results obtained from the NB classifier.

Table 2. Average hit rates with the LDA classifier using CSP channels and channels C3, C4, P3 and P4 to estimate W.

Window	Hit rate by session (% average ± standard deviation)			
	S_1	S_2	S_3	S_4
W_1	66.8 ± 10.6	68.5 ± 8.72	69.0 ± 10.3	64.8 ± 9.48
W_2	66.5 ± 10.7	67.0 ± 8.91	69.9 ± 10.6	64.5 ± 7.95
W_3	65.2 ± 9.51	66.8 ± 8.88	70.6 ± 10.1	65.1 ± 7.62
W_4	66.7 ± 9.13	66.4 ± 8.45	68.5 ± 9.91	66.8 ± 8.77
W_5	64.1 ± 10.6	67.8 ± 8.24	69.6 ± 9.01	65.4 ± 8.71
W_6	64.9 ± 9.73	66.4 ± 8.71	68.1 ± 11.2	67.4 ± 9.12

Table 3. Average hit rates with the NB classifier using all CSP channels and channels C3, C4, P3 and P4 to estimate W.

Window	Hit rate by session (% average ± standard deviation)			
	S_1	S_2	S_3	S_4
W_1	66.1 ± 10.5	66.2 ± 9.32	64.3 ± 9.76	62.9 ± 9.34
W_2	64.4 ± 10.7	64.6 ± 8.86	64.6 ± 10.6	62.4 ± 8.15
W_3	63.6 ± 9.93	65.2 ± 8.40	65.2 ± 10.6	62.1 ± 8.55
W_4	64.6 ± 11.4	63.5 ± 8.50	65.0 ± 9.87	62.0 ± 9.85
W_5	62.6 ± 10.5	65.1 ± 8.87	65.0 ± 10.4	59.7 ± 9.32
W_6	63.1 ± 10.1	64.0 ± 8.94	62.6 ± 9.55	59.5 ± 9.34

In general, no relevant variance on classification rates for differently-sized windows was observed. However, we noticed a large variability among sessions; S_3 indicated the highest hit rate. It is not possible to confirm whether the window size influenced it due to the high standard deviation found in the results; the variances among the types of window are smaller than the standard deviation of each window in the same session. Using two channels (C3 and C4) to estimate the spatial filter coefficients together with the two resulting CSP channels allowed for classification rates closer to that of using four channels (C3, C4, P3 and P4) to estimate the filter coefficients through the four resulting CSP channels. The worst result came from the tow CSP channels obtained by calculating the coefficients through all four EEG signals. These findings provide a good indication that using small windows (about 1 s) is an advantage, since it reduces the processing time, a common concern for systems tested online.

We also noticed that the increase of channels degraded the classification rates mainly for sessions S_3 and S_4. Its probable cause is the fact that only two channels (C3 and C4) showed high motor activity. Studies reporting the occurrence of high hit rates following this same methodology usually have a larger amount of channels, using 55 and 56 channels of a modified 10–20 system. Larger amounts of channels increase the resolution in the motor area, providing more discriminatory information among classes than only four channels, which probably leads to better classification rates. High variability indexes caused by high leakages indicate that classification depends both on training and testing sets, which is a common characteristic in this kind of study [36]. As a comparison, Table 4 shows the hit rate obtained only from channels C3 and C4 in session S3. Results were found by substituting matrix *W* with the identity matrix. Using an unfiltered signal allowed slightly lower rates than using a filter calculated by two EEG and two CSP channels and estimated by four EEG channels considering four CSP channels as the input characteristic. However, it is not possible to verify whether using the filter had significantly improved the classification result, because of the high variability. In order to verify whether the filter increases the EEG signal hit rate, further studies using larger amounts of channels covering the motor area and removing artifacts are necessary.

The LDA classifier again showed classification rates higher than the NB classifier. Using smaller windows significantly degraded classification, indicating that a significant amount of data is necessary for energy estimation. It is worth mentioning that the expected value of the signal energy gets closer to the real value by increasing the number of samples for its estimation [30]. Thus, smaller windows have a higher variability of values, which degrades the classification rate. Using higher sampling rates can lead to a better characterization of the signal energy in smaller windows.

Table 4. Average hit rates using channels C3 and C4 for classification.

Window	Hit rate session S_3 (% average ± standard deviation)	
	LDA	**Naive Bayes**
W_1	67.8 ± 9.75	63.2 ± 10.5
W_2	68.3 ± 8.85	63.3 ± 10.0
W_3	68.2 ± 10.4	63.7 ± 11.5
W_4	67.1 ± 9.65	63.3 ± 8.68
W_5	68.8 ± 10.2	63.1 ± 10.7
W_6	66.7 ± 9.65	61.2 ± 10.1

4.6. Classification Based on the Spectral Components as the Input Feature

This session shows the classification results obtained from spectral components. The components selected according to the higher values given by the *LI* index of the frequencies were inserted into the features vector. As a comparison, we used 2, 4 and 10 spectral components with two (C3 and C4) and four (C3, C4, P3 and P4) channels. It is worth mentioning that the amount of features is given by the product of the number of spectral components by the number of channels. Tables 5 and 6 show the results of four sessions using two and four channels, respectively.

Table 5. Average hit rates using the spectral components of two channels (C3 and C4).

Session	Spectral Components					
	2		**4**		**10**	
	LDA	**NB**	**LDA**	**NB**	**LDA**	**NB**
S_1	62.5 ± 16.6	62.9 ± 17.5	62.5 ± 13.3	65.4 ± 15.3	61.2 ± 16.6	63.5 ± 16.3
S_2	60.7 ± 14.9	52.8 ± 12.4	61.6 ± 13.6	55.7 ± 13.9	61.9 ± 14.5	56.1 ± 14.5
S_3	60.9 ± 16.0	57.3 ± 15.9	60.6 ± 15.2	56.8 ± 15.2	60.6 ± 13.5	60.3 ± 14.3
S_4	62.4 ± 14.0	59.4 ± 13.1	60.9 ± 11.8	58.8 ± 11.9	68.4 ± 11.8	65.4 ± 11.6

Table 6. Average hit rates using the spectral components of four channels (C3, C4, P3 and P4).

Session	Spectral Components					
	2		4		10	
	LDA	NB	LDA	NB	LDA	NB
S_1	62.6 ± 16.2	64.9 ± 18.4	59.6 ± 16.1	68.4 ± 17.3	54.6 ± 17.2	65.8 ± 13.5
S_2	62.6 ± 13.9	50.7 ± 13.4	64.3 ± 12.5	54.1 ± 14.9	59.3 ± 13.6	51.7 ± 14.8
S_3	56.1 ± 15.2	55.7 ± 13.3	52.7 ± 15.9	53.8 ± 16.1	45.9 ± 15.6	51.1 ± 15.9
S_4	58.9 ± 12.4	51.0 ± 14.1	60.1 ± 12.6	55.4 ± 13.0	60.8 ± 12.7	54.7 ± 11.9

Our findings indicate a lower average hit rate than those obtained through the energy of the CSP filtered signal. A higher variability was also observed, indicating irregularity in classification rates. The LDA classifier maintained a higher hit rate among sessions, except for S_1 when using four EEG channels. The number of channels had influenced mainly the hit rates given by the NB classifier, probably because the high dependence maintained by the channels' signals violate its principle of considering only independent features. The number of spectral components had not significantly influenced classification rates, which is an interesting finding, since the number of features directly influences the computational efficiency in online applications. The high variability of the data shows that using this kind of feature made the classifier very dependent on the training set, so as not to generalize data classification in a satisfactory manner.

Results indicate both variability and hit rates similar to those obtained through the signal energy and the CSP filter. These results indicate that using spectral components as input feature requires "well behaved" stationary signals, since it is usually necessary to use larger windows for a good frequency spectrum resolution, so that non-stationary behavior is not desirable. By comparing the laterality indexes over time, we can verify that the volunteers are able to maintain the movement laterality for approximately 1 s between 4 and 5 s. The problem of using much smaller windows is that the loss of frequency leads to the loss of information within the sensorimotor rhythm, since it already has activity in narrow frequency rhythms.

To confirm the investigation, ANOVA and multiple comparisons were used. For the statistical validation methodology, three-factor experiments were used (three-factor fixed effects model). In general, factorial designs are most efficient for this type of experiment. By factorial design, we mean the investigation of all possible combinations of the levels of the factors in each complete trial or replicate of the experiment. ANOVA provides a statistical test of whether or not the means of several groups are all equal. If they are not all the same, you may need information about which pairs of means are significantly different and which are not. A multiple comparison procedure is a test that can provide such information. Two means are significantly different if their intervals are disjoint and not significantly different if their intervals overlap. This experimental design is a completely randomized design. Consider the three-factor factorial experiment with underlying model Equation (23):

$$Y_{ijkl} = \mu + \tau_i + \beta_j + \gamma_k + (\tau\beta)_{ij} + (\tau\gamma)_{ik} + (\beta\gamma)_{jk} + (\tau\beta\gamma)_{ijk} + \epsilon_{ijkl} \begin{cases} i = 1,2,\dots,a \\ j = 1,2,\dots,b \\ k = 1,2,\dots,c \\ l = 1,2,\dots,n \end{cases} \tag{23}$$

where μ is the overall mean effect, τ_i is the effect of the i-th level of factor A (electrode placement), β_j is the effect of the j-th level of factor B (volunteers), γ_k is the effect of the k-th level of factor C (different sessions), $(\tau\beta)_{ij}$ is the effect of the interaction between A and B, $(\tau\gamma)_{ik}$ is the effect of the interaction between A and C, $(\beta\gamma)_{jk}$ is the effect of the interaction between B and C, $(\tau\beta\gamma)_{ijk}$ is the effect of the interaction between A, B and C and ϵ_{ijkl} is a random error component having a normal distribution with mean zero and variance σ.

An analysis for an experiment using single factor (classifiers), using analysis of variance, was also made. It was found that the iteration between the classifiers is significant, *i.e.*, they present distinct values for the database presented.

Notice that the model contains three main effects (A, B and C), three two-factor interactions, a three-factor interaction and an error term. The F-test on the main effects and interactions follows directly from the expected mean squares. These ratios follow F distributions under the respective null hypotheses. We used α = 0.05 (significance level). The analysis of variance for a three-factor experiment showed that the main effects, due to the electrode placement, volunteers and different sessions, are significant, *i.e.*, there is strong evidence to conclude that the variances of the three main effects are different.

5. Conclusions

Our findings allow us to conclude that it is possible to classify EEG signals through information about hand movement behavior. We also verified the physiological effect of hand movement in the brain through noninvasive measurements by using six EEG channels. The volunteers showed activity lateral to the movement within the μ rhythm for channels located both in central and parietal areas of the lobe. Activity for central and parietal channels within the β rhythm was also observed; however, no relevant discriminatory activity within this band was verified. The spatial filter did not significantly increase the hit rate in either classifier. This probably happened because the obtained signals did not show good spatial resolution. Besides, there were only six channels available, of which only four showed activities related to sensorimotor rhythms. Furthermore, this technique is highly sensitive to artifacts for filter estimation, and the signals were not handled so as to remove these artifacts. Table 7 presents the classification hit rate for right and left hand of other studies. Note that this studies uses other databases and different feature extraction:

Table 7. Hit rate for other studies.

Classification Method	Accuracy (%)	Reference
Gaussian Support Vector Machines	86	[48]
LDA	61	[48]
Multi-Layer Neural Network	80.4	[49]
LDA	80.6	[49]
Hidden Markov Model	81.4	[50]
Finite Impulse Neural Network	87.4	[51]
Morlet Wavelet and Bayes Quadratic integrated over time	89.3	[52]
LDA	65.6	[53]

Further studies using a cap with higher electrode resolution are required to validate the CSP filter theory. By using signal energy, with or without the CSP filter, we were able to obtain a good behavior in the classifiers and to maintain the hit rates, even with small one-second windows. The use of spectral components proved to be very sensitive to the quality of the signal and, therefore, a secondary choice for using classification features. In turn, the periodogram proved to be a useful tool to show which frequency bands are more relevant for classification. Using the data obtained through the periodogram together with other preprocessing techniques, such as principal components analysis, allows the development of algorithms for the automatic selection of features. We found lower rates primarily due to the lack of training by the volunteers and feedback, as well as the use of uncertified equipment and experimental runs in an uncontrolled environment. In order to increase the discriminatory activity between the two movement classes, results show the necessity of using feedback during the experiment, such as was performed by the BCI Competition II [28]. Although the LDA classifier had maintained higher rates than the NB classifier, they showed similar performance.

It was possible to confirm whether the use of the LDA is different than NB by the ANOVA analysis, indicating that the use of LDA can improve the classification hit rate rather than NB. Thus, it is possible to say that the statistical parameters of the volunteers are quite distinct from each other. The results of this model showed that the interactions are true once $(\tau\beta)_{ij}$, $(\tau\gamma)_{ik}$, $(\beta\gamma)_{jk}$ and $(\tau\beta\gamma)_{ijk}$ are significant.

Author Contributions

Juliano Machado developed acquisition software and the experimental protocol. Alexandre Balbinot performed the analysis of the experimental data. Both participated in other phases of the work.

Conflicts of Interest

The authors declare no conflict of interest.

References

1. Wolpaw, J.R.; Birbaumer, C.; McFarland, D.J.; Pfurtscheller, G.; Vaughan, T.M. Brain-computer interfaces for communication and control. *Clinic. Neurophysiol.* **2002**, *113*, 767–791.
2. Neuper, C.; Muller, G.R.; Kubler, A.; Birbaumer, N.; Pfurtscheller, G. Clinical application of an EEG-based brain-computer interface: A case study in a patient with severe motor impairment. *Clinic. Neurophysiol.* **2003**, *114*, 399–409.
3. Blankertz, B.; Dornhege, G.; Krauledat, M.; Muller, K.R.; Curio, G. The non-invasive Berlin brain-computer interface: Fast acquisition of effective performance in untrained subjects. *NeuroImage* **2007**, *37*, 539–550.
4. Chatterjee, A.; Aggarwal, V.; Ramos, A.; Acharya, S.; Thakor, N.V. A brain-computer interface with vibrotactile biofeedback for haptic information. *J. Neuroeng. Rehabil.* **2007**, *4*, 4–40.
5. Kamousi, B.; Amini, A.N.; He, B. Classification of motor imagery by means of cortical current density estimation and Von Neumann entropy for brain-computer interface applications. *J. Neural Eng.* **2007**, *4*, 17–25.

6. Pfurtscheller, G.; Brunner, C.; Schlogl, A.; Lopes da Silva, F.H. Mu rhythm (de)synchronization and EEG single-trial classification of different motor imagery tasks. *Neuroimage* **2006**, *31*, 153–159.

7. Pineda, J.A.; Silverman, D.S.; Vankov, A.; Hestenes, J. Learning to control brain rhythms: Making a brain-computer interface possible. *IEEE Trans. Neural Syst. Rehabil. Eng.* **2003**, *11*, 181–184.

8. Birbaumer, N.; Ghanayim, N.; Hinterberger, T.; Iversen, I.; Kotchoubey, B.; Kubler, A. A spelling device for the paralysed. *Nature* **1999**, *398*, 297–298.

9. Bayliss, J.D. Use of the evoked potential P3 component for control in a virtual apartment. *IEEE Trans. Neural Syst. Rehabil. Eng.* **2003**, *11*, 113–116.

10. Hoffmann, U.; Vesin, J.M.; Ebrahimi, T.; Diserens, K. An efficient P300-based brain-computer interface for disabled subjects. *J. Neurosci. Methods* **2008**, *167*, 115–125.

11. Lalor, E.C.; Kelly, S.P.; Finucane, C.; Burke, R.; Reilly, R.B. Steady-state VEP-based brain-computer interface control in an immersive 3D gaming environment. *EURASIP J. Appl. Signal Process.* **2005**, *2005*, 3156–3164.

12. Middendorf, M.; McMillan, G.; Calhoun, G.; Jones, K.S. Brain-computer interfaces based on the steady-state visual-evoked response. *IEEE Trans. Rehabil. Eng.* **2000**, *8*, 211–214.

13. Galan, F.; Nuttin, M.; Lew, E.; Ferrez, P.W.; Vanacker, G.; Philips, E. A brain-actuated wheelchair: Asynchronous and non-invasive brain-computer interfaces for continuous control of robots. *Clinic. Neurophysiol.* **2008**, *119*, 2159–2169.

14. Pfurtscheller, G. Induced oscillations in the alpha band: Functional meaning. *Epilepsia* **2003**, *44*, 2–8.

15. Decety, J.; Inqvar, D. Brain structures participating in mental simulation of motor behavior: A neuropsychological interpretation. *Acta Psychol.* **1990**, *73*, 13–34.

16. Jeannerod, M.; Frak, V. Mental imaging of motor activity in humans. *Curr. Opin. Neurobiol.* **1999**, *9*, 735–739.

17. McFarland, D.J.; Wolpaw, J.R. Brain-computer interface operation of robotic and prosthetic devices. *IEEE Comput. Soc.* **2008**, *41*, 82–86.

18. Pfurtscheller, G. Functional topography during sensorimotor activation studied with event-related desynchronization mapping. *J. Clinic. Neurophysiol.* **1989**, *6*, 75–84.

19. Neuper, C.; Pfurtschelle, G. Event-related negativity and alpha band desynchronization in motor reactions. *EEG EMG Z Elektroenzephalogr. Elektromyogr. Verwandte Geb.* **1992**, *2*, 55–61.

20. Pfurtscheller, G.; Neuper, C. Motor imagery activates primary sensorimotor area in humans. *Neurosci. Lett.* **1997**, *239*, 65–68.

21. Caldara, R.; Deiber, M.P.; Andrey, C.; Michel, C.; Thut, G.; Hauert, C.A. Actual and mental motor preparation and execution: A spatiotemporal ERP study. *Exp. Brain Res.* **2004**, *159*, 389–399.

22. Pfurtscheller, G.; Zalaudek, K.; Neuper, C. Event-related beta synchronization after wrist, finger and thumb movement. *Electroencephalogr. Clinic. Neurophysiol.* **1998**, *2*, 154–160.

23. Neuper, C.; Pfurtscheller, G. Evidence for distinct beta resonance frequencies in human EEG related to specific sensorimotor cortical areas. *Clinic. Neurophysiol.* **2001**, *11*, 2084–2097.

24. Stancak, A., Jr.; Feige, B.; Lucking, C.H.; Kristeva-Feige, R. Oscillatory cortical activity and movement-related potentials in proximal and distal movements. *Clinic. Neurophysiol.* **2000**, *4*, 636–650.

25. Cassim, F.; Monaca, C.; Szurhaj, W.; Bourriez, J.L.; Defebvre, L.; Derambure, P.; Guieu, J.D. Does post-movement beta synchronization reflect and idling motor cortex? *Neuroreport* **2001**, *17*, 3859–3863.

26. Gernot, R.M.P.; Zimmerman, D.; Graimann, B.; Nestinger, K.; Korisek, G.; Pfurtscheller, G. Event-related beta EEG-changes during passive and attempted foot movements in paraplegic patients. *Brain Res.* **2007**, *1137*, 84–91.

27. Nam, C.S.; Jeon, Y.; Kim, Y.J.; Lee, I.; Park, K. Movement imagery-related lateralization of event-related de(synchronization) (ERD/ERS): Motor-imagery duration effects. *Clinic. Neurophysiol.* **2011**, *3*, 567–577.

28. BCI Competition II. Available online: http://www.bbci.de/competition/ii/ (accessed on 12 July 2011).

29. Herman, P.; Prasad, G.; McGinnity, T.M.; Coyle, D. Comparative analysis of spectral approaches to feature extraction for EEG-based motor imagery classification. *IEEE Trans. Neural Syst. Rehabil. Eng.* **2008**, *4*, 317–326.

30. Bendat, J.; Piersol, A. *Random Data Analysis: Analysis and Measurement Procedures;* John Wiley: New York, NY, USA, 1986.

31. Dornhege, G.; Millán, J.R.; Hinterberger, T.; McFarland, D.; Muller, K.R. *Towards Brain-Computer Interfacing*; The MIT Press: Cambridge, MA, USA, 2007.

32. Oppenheim, A.; Schaefer, R.; Buck, J. *Discrete-Time Signal Processing*; Prentice Hall: Upper Saddle River, NJ, USA, 1998.

33. Stoica, P.; Moses, R. *Spectral Analysis of Signals*; Prentice Hall: Upper Saddle River, NJ, USA, 2005.

34. Welch, P.D. The use of Fats Fourier Transform for the estimation of power spectra: A method based on time averaging over shot, modified periodograms. *IEEE Trans. Audio Electroacoust.* **1967**, *2*, 70–73.

35. Pfurtscheller, G.; Neuper, C.; Brunner, C.; Lopes da Silva, F. Beta rebound after different types of motor imagery in man. *Neurosci. Lett.* **2005**, *378*, 156–159.

36. Ramoser, H.; Müller-Gerking, J.; Pfurtscheller, G. Optimal spatial filtering of single trial EEG during imagined hand movement. *IEEE Trans. Rehabil. Eng.* **1998**, *8*, 441–446.

37. Mitchel, T. *Machine Learning*; McGraw-Hill Science: New York, NY, USA, 1997.

38. Fukunaga, K. *Introduction to Statistical Pattern Recognition*; Academic Press: San Diego, CA, USA, 1990.

39. Duda, R.O.; Hart, P.E.; Stork, D.G. *Pattern Classification*; Wiley-Interscience: Toronto, Canada, 2000.

40. Jasper, H.H. The ten twenty electrode system. *Int. Fed. Electroencephalogr. Clinic. Neurophysiol.* **1958**, *10*, 371–375.

41. Sanei, S.; Chambers, J.A. *EEG Signal Processing*; John Wiley & Sons Ltd.: Chichester, UK, 2007.

42. PfurtschNeller, G.; Lopes da Silva, F.H. Event-related EEG/MEG synchronization and desynchronization: Basic principles. *Clinic. Neurophysiol.* **1999**, *11*, 1842–1857.

43. Carra, M.; Balbinot, A. Evaluation of sensorimotor rhythms to control a wheelchair. In Proceedings of the 2013 ISSNIP Biosignals and Biorobotics Conference (BRC2013), Rio de Janeiro, Brazil, 18–20 February 2013; pp. 1–4.

44. Doyle, L.M.F.; Yarrow, K.; Brown, P. Lateralization of event-related beta desynchronization in the EEG during pre-cued reaction time tasks. *Clinic. Neurophysiol.* **2005**, *116*, 1879–1888.

45. Bhattacharyya, S.; Khasnobish, A.; Amit, K.; Tibarewala, D.N.; Nagar, A.K.; Irvine, D.; Gongora, M. Performance analysis of left/right hand movement classification from EEG signal by intelligent algorithms. In Proceedings of 2011 IEEE Symposium on Computational Intelligence, Cognitive Algorithms, Mind and Brain (CCMB), Paris, France, 11–15 April 2011; pp.1–8.

46. Carra, M.; Balbinot, A. Development of a brain-computer interface system based on sensorimotor rhythms. In Proceedings of the 2012 ISSNIP Biosignals and Biorobotics Conference (BRC2012), Manaus, Brazil, 9–11 January 2012; pp. 21–25.

47. Garcia, G.N.; Ebrahimi, T.; Vesin, J.M. Support vector EEG classification in the Fourier and time-frequency correlation domains. In Proceedings of 1st International IEEE EMBS Conference on Neural Engineering, Capri Island, Italy, 20–22 March 2003; pp. 591–594.

48. Hoffmann, U.; Garcia, G.; Vesin, J.M.; Diserens, K.; Ebrahimi, T. A boosting approach to p300 detection with application to brain–computer interfaces. In Proceedings of 2nd International IEEE EMBS Conference Neural Engineering, Arligton, TX, USA, 16–19 March 2005; pp. 97–100.

49. Boostani, R.; Moradi, M.H. A new approach in the BCI research based on fractal dimension as feature and Adaboost as classifier. *J. Neural Eng.* **2004**, *1*, 51–56.

50. Obermeier, B.; Guger, C.; Neuper, C.; Pfurtscheller, G. Hidden Markov models for online classification of single trial EEG data. *Pattern Recognit. Lett.* **2001**, *22*, 1299–1309.

51. Haselsteiner, E.; Pfurtscheller, G. Using time-dependant neural networks for EEG classification. *IEEE Trans.Rehabil. Eng.* **2000**, *8*, 457–463.

52. Lemm, S.; Schafer, C.; Curio, G. BCI competition 2003–data set III: Probabilistic modeling of sensorimotor mu rhythms for classification of imaginary hand movements. *IEEE Trans. Biomed. Eng.* **2004**, *51*, 1077–1080.

53. Solhjoo, S.; Moradi, M.H. Mental task recognition: A comparison between some of classification methods. In Proceedings of BIOSIGNAL 2004 17th International EURASIP, Brno, Czech Republic, 23–24 June 2004; pp. 273–277.

12

A Novel Transdermal Power Transfer Device for the Application of Implantable Microsystems

Jing-Quan Liu *, Yue-Feng Rui, Xiao-Yang Kang, Bin Yang, Xiang Chen and Chun-Sheng Yang

National Key Laboratory of Science and Technology on Micro/Nano Fabrication, Key Laboratory of Shanghai Education Commission for Intelligent Interaction and Cognitive Engineering, Department of Micro/Nano-electronics, Shanghai Jiao Tong University, Shanghai 200240, China; E-Mails: yfrui@sjtu.edu.cn (Y.-F.R.); xykang@sjtu.edu.cn (X.-Y.K.); binyang@sjtu.edu.cn (B.Y.); xiangchen@sjtu.edu.cn (X.C.); csyang@sjtu.edu.cn (C.-S.Y.)

* Author to whom correspondence should be addressed; E-Mail: jqliu@sjtu.edu.cn

Academic Editor: Paul Ronney

Abstract: This paper presents a transdermal power transfer device for the application of implantable devices or systems. The device mainly consists of plug and socket. The power transfer process can be started after inserting the plug into the socket with an applied potential on the plug. In order to improve the maneuverability and reliability of device during power transfer process, the metal net with mesh structure were added as a part of the socket to serve as intermediate electrical connection layer. The socket was encapsulated by polydimethylsiloxane (PDMS) with good biocompatibility and flexibility. Two stainless steel hollow needles placed in the same plane acted as the insertion part of the needle plug, and Parylene C thin films were deposited on needles to serve as insulation layers. At last, the properties of the transdermal power transfer device were tested. The average contact resistance between needle and metal mesh was 0.454 Ω after 50 random insertions, which showed good electrical connection. After NiMH (nickel-metal hydride) batteries were recharged for 10 min with current up to 200 mA, the caused resistive heat was less than 0.6 °C, which also demonstrated the low charging temperature and was suitable for charging implantable devices.

Keywords: implantable; transdermal; power transfer; PDMS; Parylene

1. Introduction

Implantable microsystems are widely used in biomedical engineering for diagnostics and monitoring [1–3], drug delivery [4–7], artificial prosthesis [8–11], and so on. A typical implantable microsystem usually contains three parts. The first part is a sensor or actuator unit used for recording signals or stimulating nerve cells. The second one is a control unit used for signal processing and generating. And the last one is a power unit used for supplying power.

Figure 1a shows conceptual diagram of the typical implantable microsystem [12]. The power supply unit plays a very important role in implantable microsystems due to its effect on the lifetime of the systems [13–16]. Nowadays, there are two kinds of power supply units widely applied in implantable microsystems, which are primary batteries [17] and rechargeable batteries [18–21]. As to long-term implantation of implantable microsystems for several decades, primary batteries are limited due to their limited battery capacity [22–26]. In cardiac pacing systems, heart failure might cause casualties due to the depletion of the batteries before replacement [27–29]. However, the replacement of batteries will increase the cost, the risk and suffering of patients. Rechargeable batteries for microsystems provide a good alternative to solve this problem. For non-invasive purposes, most implantable microsystems are recharged by wireless power transfer method [18–20,30–33]. However, the wireless recharging method with relatively low power transmission efficiency will lead to a very long charging time. Meanwhile, the electromagnetic radiation produced by wireless power transfer might cause skin and tissue injury, even the potential risk of cancer. As to high power consumption systems, transdermal power transfer method has some advantages, such as micro-invasive characteristics, high power transmission efficiency and low charging temperature [21,32,34–39]. Transdermal power transfer devices charged by small needles, which could lead to a small injury compared with wireless power transfer device, because electromagnetic radiation injury produced by wireless power transfer is much worse than the temperature increase; the literature mentioned above did not talk about the effects of the temperature increase.

Although the transdermal recharging method has the risk of causing skin infection during charging process, this risk can be eliminated by sterilizing the insertion needle as well as the injection of saline. Furthermore, several hours of power transfer by transdermal recharging can keep an implantable system working for several years. At present, the emerged transdermal recharging device usually contains two parts. One part is the plug used for the connection with the external power supply. The other one is the socket implanted *in vivo* used for the connection with the internal power supply. The socket with two metal contact-springs acts as an intermediate electrical connection layers, which might lead to poor maneuverability due to its misalignment after implantation *in vivo*. The plug is fabricated by coating Parylene C thin film on a hollow metal needle. However, due to its double-layer structure of metal contact-springs, the coating might be scratched during the insertion process.

In this paper, a novel type of transdermal power transfer device for implantable device application is presented. Figure 1b shows the schematic diagram of the proposed transdermal power transfer system. The socket with two pieces of metal nets placed in the same plane was encapsulated by polydimethylsiloxane (PDMS), realizing good maneuverability, reliable electrical connection and thickness reduction of the device. The needle part of the plug was fabricated by two hollow metal needles coated with Parylene C thin films to serve as an insulation layer. Finally, the electrical connection, waterproofness, penetrating force, recharging temperature and effectiveness of device were tested, respectively.

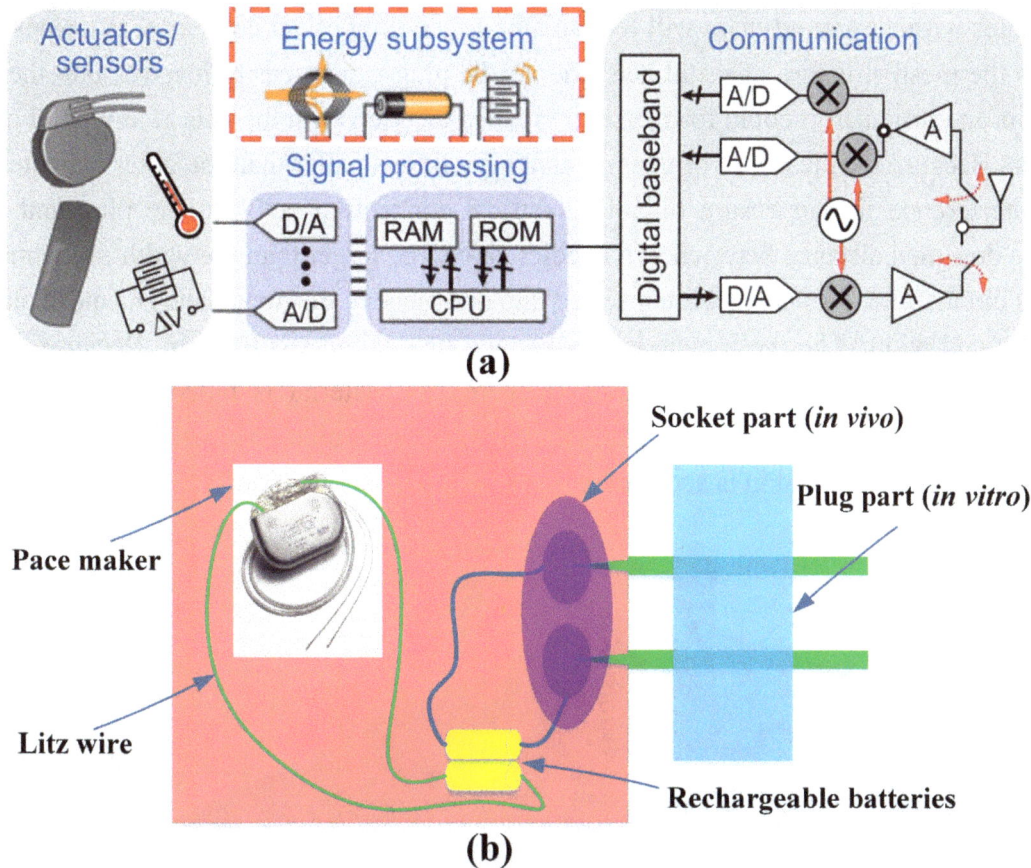

Figure 1. (**a**) The conceptual diagram of a typical implantable microsystem and (**b**) the schematic diagram of the proposed transdermal power transfer system.

2. Experimental Section

2.1. Design

A device with the following properties was desired: easy recharging, good biocompatibility, waterproof, non- or micro-invasive, low charging temperature, *etc.* PDMS and Parylene C have good biocompatibility and flexibility and have been widely used in implantable microsystems for encapsulation [40–42]. Figure 2 shows the schematic diagram of the transdermal power transfer device. Combined with Figure 1b, we chose litz wire with 38 AWG (American Wire Gage) and the designed maximum current was 500 mA. Since the devices were well encapsulated, we used the DC current for charging after a careful examination. This is a suitable, safe recharging method for implantable applications. By using the interference fit between needles and stainless steel nets, reliable electrical connection and good maneuverability during power transfer process was obtained. The power transfer process can be started after inserting the plug into the socket with an applied recharging potential on the plug as shown in Figure 2a. Two pieces of stainless steel net, for the sockets, were placed in the same plane and used for connecting to the cathode and anode of the internal power supply, respectively.

The socket was encapsulated in a PDMS encapsulation shell as shown in Figure 2b. The plug was fabricated by the installation of two stainless steel hollow needles in the plastic plane. A Parylene C thin film used as insulation layer was chemical vapor deposited (CVD) on needles. To decrease the charging temperature, the plug was designed with a notch on the plastic plane as shown in Figure 2c. The notch part of the needles with air around them will lead to a better cooling effect compared to encapsulate in PDMS. Due to the mesh structure of metal nets, the needle plug can be easily inserted into the socket through the skin. In particular, it could form an electrical connection with the nets at any point over the connection area. Because the mesh size of this metal net is a little smaller than the outer diameter of the needles, the interference fit can ensure reliable electrical connection between the plug and socket. Meanwhile, the designed distance between two needles should be larger than the width of the metal net to avoid a short circuit when two needles are inserted into one piece of the metal net. The mesh plates are 0.5 cm in length and width. The mesh plate is flexible and its thickness is 0.2 mm. Because the mesh plate is encapsulated in a PDMS encapsulation shell and its material is biocompatible, it can be implanted permanently. The penetration of the metal needles is similar to a needle injection. Thus, the resulting pain or infection of skin is acceptable, same as a needle injection.

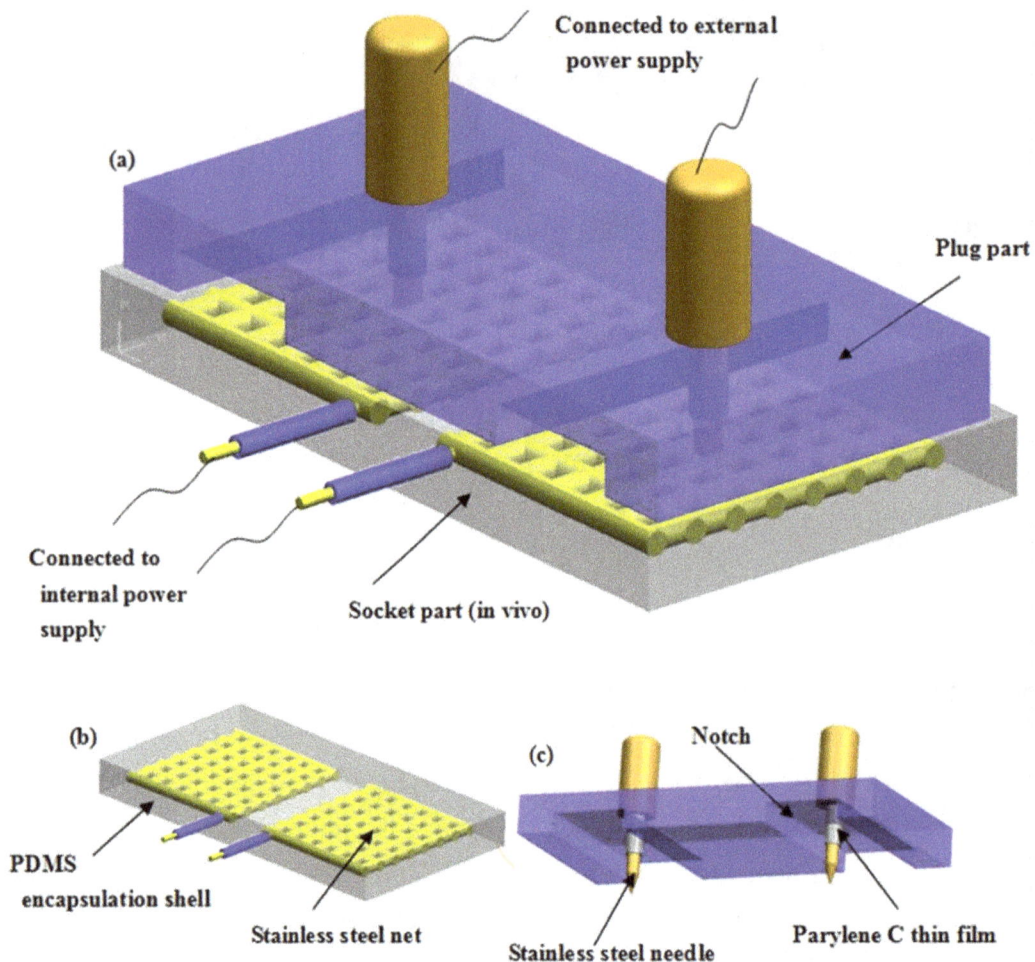

Figure 2. Conceptual diagram of the typical implantable microsystem. The schematic of the transdermal power transfer device. (**a**) The schematic of transdermal power transfer process. The process can be started after the insertion of the plug into the socket then applying a certain recharging potential to the plug. (**b**) The schematic of socket part. The stainless steel nets, with mesh size of 0.5 mm, match with the needle plug and form a good electrical connection. The whole plug was encapsulated by polydimethylsiloxane (PDMS) by cast method. (**c**) The schematic of plug. Parylene C thin film (5 μm) was chemical vapor deposited on needles for insulation and the notch on plastic plane was designed to decrease the recharging temperature.

2.2. Fabrication

Figure 3a shows the fabrication process of the socket. First, the PDMS (Sylgard® 184, Dow Corning, Midland, MI, USA) mixture (Base:Curing Agent = 10:1) was prepared. Then the PDMS packaging shell without cover was fabricated using a casting method with a curing process at the temperature of 75 °C for 3 h. The thickness of the PDMS film used for device packaging is 0.15 cm. The length, width and the height of the cuboid shell were 20 mm, 10 mm and 3 mm, respectively.

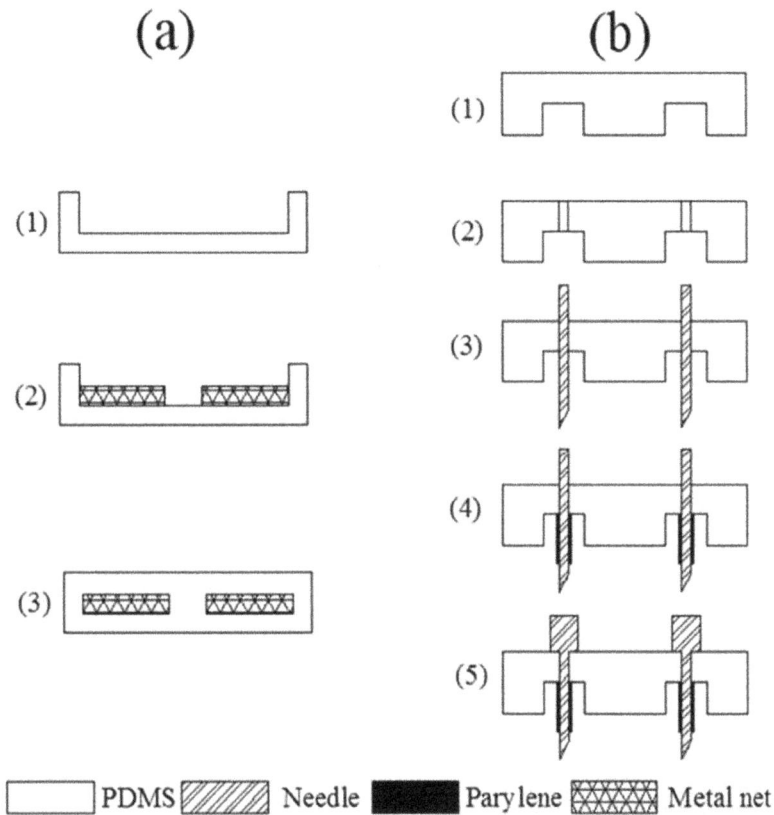

Figure 3. The fabrication process of device. (**a**) The fabrication process of the plug. (**b**) The fabrication process of socket.

Second, the two pieces of stainless steel net, with a mesh size of 0.5 mm, were connected to lead wires. After that, they were placed in the PDMS shell in the same plane. Last, the PDMS cover was casted to create a whole-packaging shell. Figure 3b shows the fabrication process of the plug. First, a plastic plane with notches was prepared. Second, two holes with diameter of 0.6 mm were drilled on the notch position for the installation of needles. Third, the needles with diameter of 0.55 mm were installed in plastic plane and stuck with glue. Fourth, 5 μm of Parylene C thin film was chemical vapor deposited (Parylene deposition system (PDS) 2010, Specialty Coating Systems (SCS), Indianapolis, IN, USA) on the needles as an insulation layer. The tips of the needles are rubbed on a friction plate to remove the Parylene. Last, the needles were connected to the metal plug for easy connection to an external power supply. Figure 4a,b shows the fabricated socket and plug.

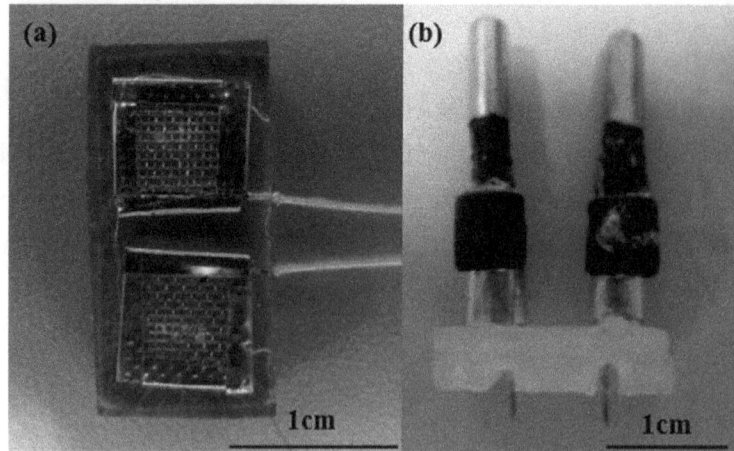

Figure 4. The fabricated device. (**a**) The fabricated socket with two pieces of stainless steel net inside. (**b**) The fabricated plug with two needles.

3. Results and Discussion

3.1. Electrical Connection

The transdermal power transfer device with good reliability can result in stable power transfer. Good maneuverability of the device also affects its practicality. Because of the interference fit between stainless steel nets (mesh size = 0.5 mm) and needle plug (diameter = 0.55 mm), the good electrical connection could be formed. In order to evaluate electrical connection reliability and maneuverability, the contact resistances between needle and stainless steel net were investigated with 50 random insertions. Figure 5 shows the contact resistances between single needle and stainless steel net over 50 random insertions. The resistance varies from 0.26 to 0.62 Ω. The average contact resistance of 0.454 Ω can be calculated from the data. Figure 6a,b shows the status of PDMS encapsulation shell and stainless steel net after 50 times random insertions.

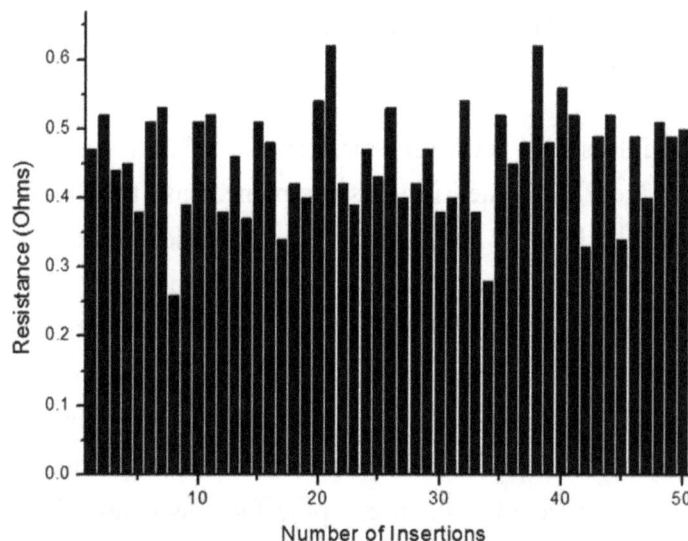

Figure 5. The contact resistances between single needle and stainless steel over 50 random insertions. The resistance varies from 0.26 to 0.62 Ω. The average contact resistance of 0.454 Ω can be calculated from the data.

Figure 6. The pictures of PDMS (**a**) and stainless steel net (**b**) after 50 random insertions.

3.2. Waterproofness

Implantable devices with good waterproofness would prevent the infiltration of tissue fluid for inner structure protection. Implantable microsystems are recharged many times after implantation and during each recharging process the PDMS encapsulation shell could be punctured again, which would increase the possibility of the infiltration of tissue fluid. The thickness of PDMS film used for waterproofness test is same as the device package. The waterproofness test was taken by injecting 0.1 M AgNO₃ solution into a cylindrical PDMS packaging shell and then dipping it into 0.9% NaCl solution for 10 months at room temperature. Figure 7 shows the status of the PDMS encapsulation shell after dipping into 0.9% NaCl solution for 10 months. Before testing, a PDMS packaging shell with nine punctures was prepared to simulate the real situation. A white precipitate of silver nitrate (AgCl) would be produced if the NaCl solution infiltrated the PDMS encapsulation shell. The chemical change can be explained by the following equation:

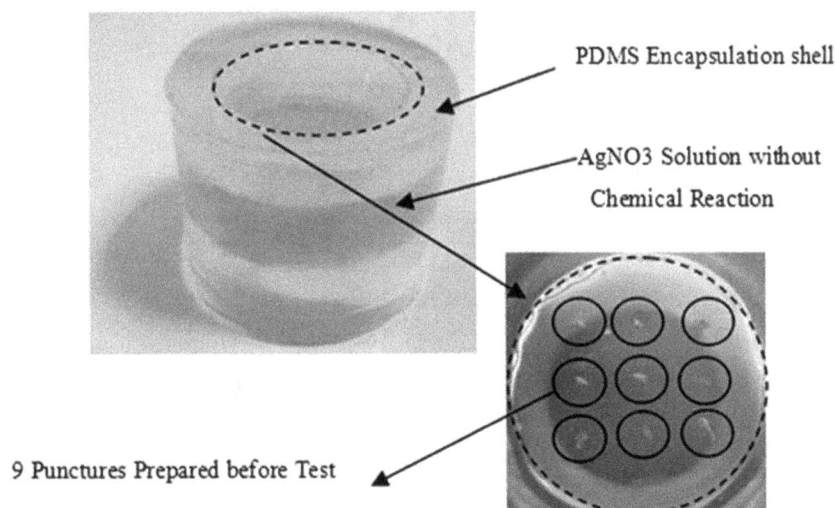

$$AgNO_3 + NaCl = AgCl\downarrow + NaNO_3 \tag{1}$$

Figure 7. The status of the PDMS encapsulation shell after dipping into 0.9% NaCl solution for five months. After five months of dipping, there was no white precipitate generated. Nine punctures (solid line circles) were prepared before the test to simulate the real situation.

As can be seen from Figure 7, there were no new chemical substances produced. The experimental results showed good waterproofness of the PDMS encapsulation shell, because of its excellent flexibility over a short time. The test will continue for a long time to evaluate long-term waterproofness of this encapsulation shell.

3.3. Penetrating Force

For easy insertion, a low penetrating force is desired. The penetrating force of the device can be calculated from a single needle's penetrating force, which is measured by a force measurement instrument. Figure 8 shows the relationship between force and displacement. It can be seen that the force is increased along with the displacement. In the PDMS region (0 < displacement < 2 mm), the penetrating force increased linearly with the displacement. The drag coefficient can be calculated as 0.65 N/mm from this curve. When the needle plug is inserted into the steel nets (2 mm < displacement < 2.5 mm), the force obviously increases and arrived at about 1.2 N. The reason is the increasing friction force caused by the interference in the fit between needle plug and nets. Then the total penetrating force of the device with two needles can be calculated by following equation:

$$F = 2(at + b) \qquad (2)$$

where F and t represent the total penetrating force and the total thickness of PDMS, respectively. a and b are the drag coefficient and interference fit force, respectively. Because a is 0.65 N/mm, b is 1.2 N, and t is 3 mm (the thickness of the designed of socket = 3 mm), the maximum penetrating force can be calculated as about 6 N.

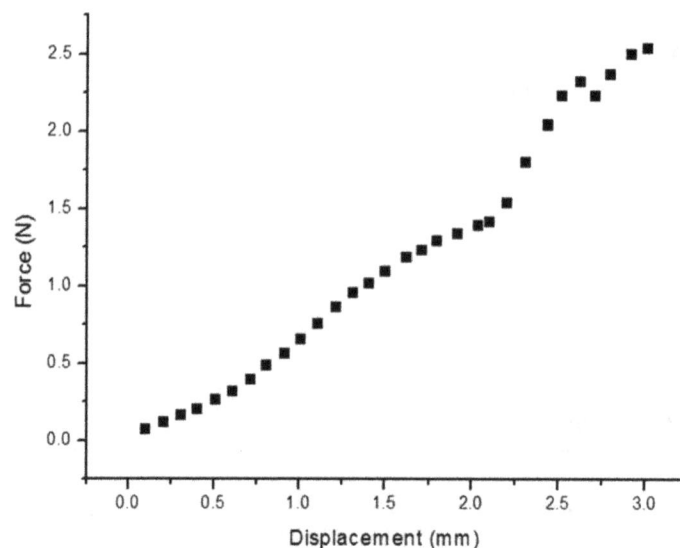

Figure 8. The relationship between force and displacement.

3.4. Charging Temperature

High charging temperature will burn the skin and tissue during power transfer process. In order to measure the charging temperature, the 1.2 V NiMH AA battery was recharged by different current levels at room temperature under a constant current power supply. Then, the temperature of the needles were measured by two thermometers at the same time. Figure 9 shows the temperature changes of the two needles under different recharging currents. The device was fully immersed in 0.9% NaCl solution with a base temperature of 26.8 °C under different recharging current levels. Since the electrical nets were encapsulated in PDMS, the dissipating speed of heat is slower compared to the notch part of the needles. This is the main reason for the rapid temperature changes under large charging currents. Thus, the design of the notch part of the needles is helping to cool the needles with the air around them. Normally, this has little effect when the charging currents is between 20 and 100 mA. However, it could help reduce the temperature change by 0.1 and 0.3 °C when the currents wee 200 and 500 mA compared to the non-notched devices, respectively. It can be seen that the charging current between 20 and 200 mA makes little temperature change. When the charging current increased to 500 mA, a dramatic increase of temperature occurred. This suggested that 200 mA recharging current would be recommended during power transfer process. It took 2.1 h to fully charge the battery under a 200 mA charging current. The local temperature rose about 0.56 °C after a full charge. Practically, the charging process could be divided into several sections to reduce the temperature changes.

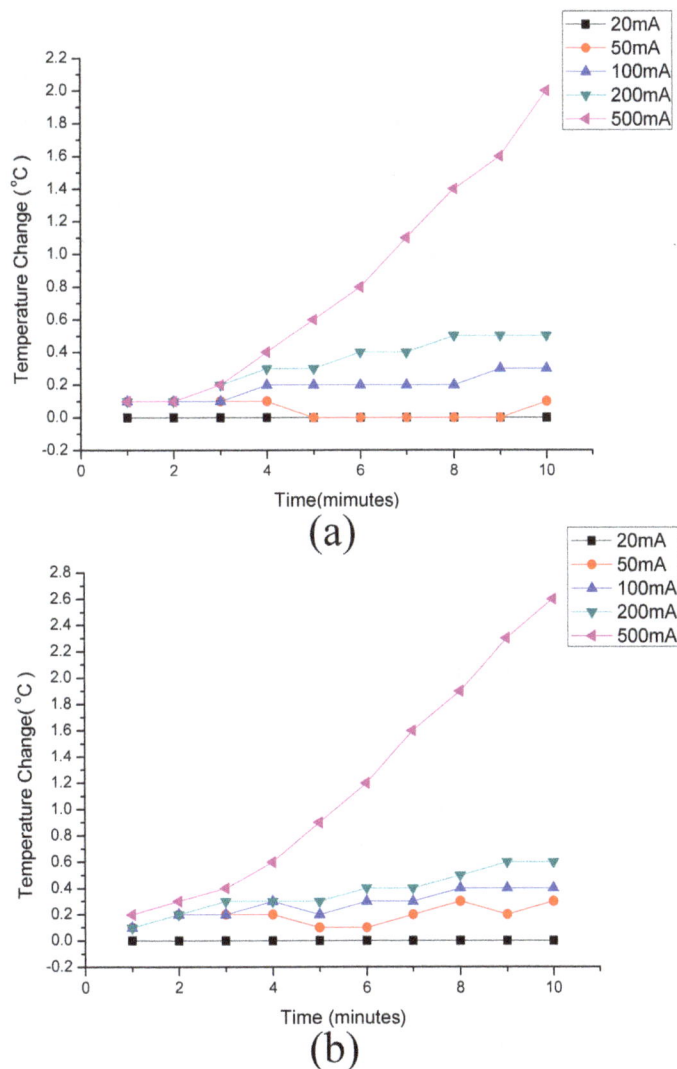

Figure 9. The temperature changes of two needles under different recharging current levels with base a temperature of 26.8 °C. (**a**) The temperature changes of Needle one. (**b**) The temperature changes of Needle two.

3.5. Implantation of Device

Figure 10 shows the implantation of device in pork tissue. Two green litz wires were connected with the anode and cathode, respectively. When the plug is inserted into the socket through the skin, the anode and cathode of the battery contacts the terminals of the plug. In the experiment, a 1.2 V NiMH AA battery was used to connect the green lead wires. Then, the multimeter was used to measure the potential from the terminals after the plug is inserted into the socket. The measured 1.2 V potential demonstrated effectiveness of electrical connection of the device in implantation occasions. Figure 10b shows the picture of the socket implanted in pork tissue by ultrasound imaging.

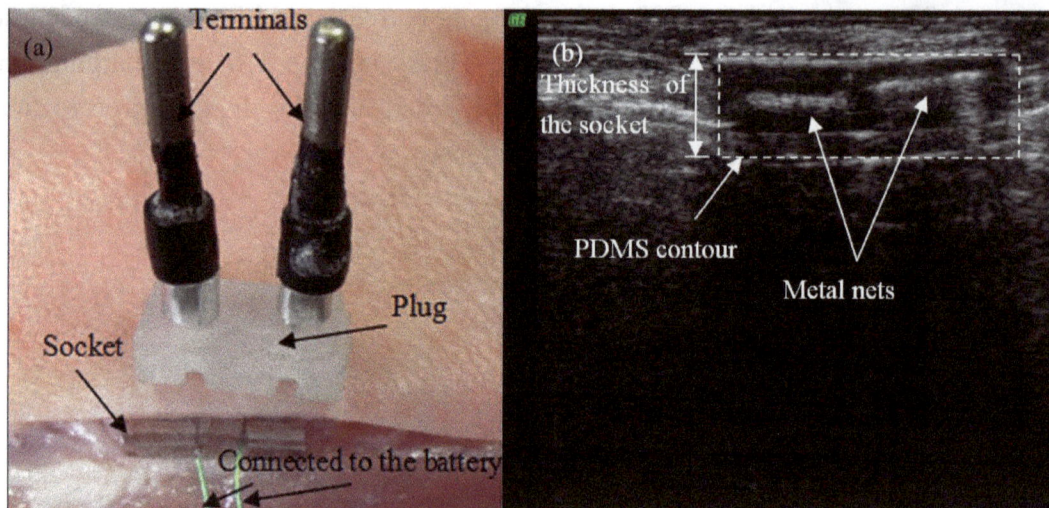

Figure 10. The implantation of device in pork tissue. (**a**) The picture of the device implanted in pork tissue. (**b**) The picture of the device implanted in pork tissue by ultrasound imaging.

4. Conclusions

In this paper, a transdermal power transfer device for implantable applications was successfully fabricated. The interference fit between needles and stainless steel nets realized the reliable electrical connection and good maneuverability during power transfer process. PDMS encapsulation shell with a number of punctures on it showed good waterproofness for preventing tissue fluid infiltration. Furthermore, the low insertion force demonstrated easy insertion of the needle plug. Moreover, the little changes of charging temperature under 200 mA also indicated safe recharging. These characteristics were all desired in charging process. This transdermal power recharging system will be suitable for implantable applications.

Acknowledgments

This work is partly supported by the National Natural Science Foundation of China (No. 51475307, 61176104), 973 Program (2013CB329401), Shanghai Municipal Science and Technology Commission (No.13511500200), Specialized Research Fund for the Doctoral Program of Higher Education (20130073110087), National Defense Pre-Research Foundation of China (No. 9140A26060313JW3385), Human Factor Key Lab (HF2012-k-01), SJTU-Funding (YG2012MS51). The authors are also grateful to the colleagues for their essential contribution to this work.

Author Contributions

Jing-Quan Liu designed the devices and the experiments, wrote large parts of the manuscript, authored almost all of the illustrations and edited the article as a whole. Yue-Feng Rui performed the microfabrication, Xiao-Yang Kang was responsible for designing and performing the electrochemical measurements, Bin Yang helped the random insertion experiment, Xiang Chen helped PDMS encapsulation shell and Chun-Sheng Yang helped the implantation of device in pork tissue.

Conflicts of Interest

The authors declare no conflict of interest.

References

1. Gamini, D.S.; Shastry, P.N. Design and measurements of implantable chip radiator and external receptor for wireless blood pressure monitoring system. In Proceedings of IEEE 2009 MTT-S International Microwave Symposium Digest, Boston, MA, USA, 7–12 June 2009; pp. 1681–1684.
2. Hao, S.Y.; Taylor, J.; Miles, A.W.; Bowen, C.R. An implantable electronic system for *in-vivo* stability evaluation of prosthesis in total hip and knee arthroplasty. In Proceedings of IEEE 2009 Instrumentation and Measurement Technology Conference, Singapore, 5–7 May 2009; pp. 167–172.
3. Ko, W.H.; Guo, J.; Ye, X.S.; Zhang, R.; Young, D.J.; Megerian, C.A. Mems acoustic sensors for totally implantable hearing aid systems. In Proceedings of IEEE 2008 International Symposium on Circuits and Systems (ISCAS 2008), Seattle, WA, USA, 18–21 May 2008; pp. 1812–1817.
4. Rahimi, S.; Sarraf, E.H.; Wong, G.K.; Takahata, K. Implantable drug delivery device using frequency-controlled wireless hydrogel microvalves. *Biomed. Microdevices* **2011**, *13*, 267–277.
5. Winzenburg, G.; Schmidt, C.; Fuchs, S.; Kissel, T. Biodegradable polymers and their potential use in parenteral veterinary drug delivery systems. *Adv. Drug Deliv. Rev.* **2004**, *56*, 1453–1466.
6. Chen, J.; Chu, M.; Koulajian, K.; Wu, X.Y.; Giacca, A.; Sun, Y. A monolithic polymeric microdevice for pH-responsive drug delivery. *Biomed. Microdevices* **2009**, *11*, 1251–1257.
7. Grider, J.S.; Brown, R.E.; Colclough, G.W. Perioperative management of patients with an intrathecal drug delivery system for chronic pain. *Anesth. Analg.* **2008**, *107*, 1393–1396.
8. Meacham, K.W.; Giuly, R.J.; Guo, L.; Hochman, S.; DeWeerth, S.P. A lithographically-patterned, elastic multi-electrode array for surface stimulation of the spinal cord. *Biomed. Microdevices* **2008**, *10*, 259–269.
9. McCreery, D.B. Cochlear nucleus auditory prostheses. *Hear. Res.* **2008**, *242*, 64–73.
10. Fayad, J.N.; Otto, S.R.; Shannon, R.V.; Brackmann, D.E. Cochlear and brainstem auditory prostheses "neural interface for hearing restoration: Cochlear and brain stem implants". *Proc. IEEE* **2008**, *96*, 1085–1095.
11. Koo, K.I.; Lee, S.; Bae, S.H.; Seo, J.M.; Chung, H.; Cho, D.I. Arrowhead-shaped microelectrodes fabricated on a flexible substrate for enhancing the spherical conformity of retinal prostheses. *J. Microelectromech. Syst.* **2011**, *20*, 251–259.
12. Chandrakasan, A.P.; Verma, N.; Daly, D.C. Ultralow-power electronics for biomedical applications. *Annu. Rev. Biomed. Eng.* **2008**, *10*, 247–274.

13. Zhang, J.; Suo, Y.M.; Mitra, S.; Chin, S.; Hsiao, S.; Yazicioglu, R.F.; Tran, T.D.; Etienne-Cummings, R. An efficient and compact compressed sensing microsystem for implantable neural recordings. *IEEE Trans. Biomed. Circuits Syst.* **2014**, *8*, 485–496.

14. Park, S.; Borton, D.A.; Kang, M.Y.; Nurmikko, A.V.; Song, Y.K. An implantable neural sensing microsystem with fiber-optic data transmission and power delivery. *Sensors* **2013**, *13*, 6014–6031.

15. Chang, C.W.; Chiou, J.C. A wireless and batteryless microsystem with implantable grid electrode/3-dimensional probe array for ecog and extracellular neural recording in rats. *Sensors* **2013**, *13*, 4624–4639.

16. Cheong, J.H.; Ng, S.S.Y.; Liu, X.; Xue, R.F.; Lim, H.J.; Khannur, P.B.; Chan, K.L.; Lee, A.A.; Kang, K.; Lim, L.S.; *et al.* An inductively powered implantable blood flow sensor microsystem for vascular grafts. *IEEE Trans. BioMed. Eng.* **2012**, *59*, 2466–2475.

17. Mallela, V.S.; Ilankumaran, V.; Rao, N.S. Trends in cardiac pacemaker batteries. *Indian Pacing Electrophysiol. J.* **2004**, *4*, 201–212.

18. Cong, P.; Suster, M.A.; Chaimanonart, N.; Young, D.J. Wireless power recharging for implantable bladder pressure sensor. In Proceedings of 2009 IEEE Sensors, Christchurch, New Zealand, 25–28 October 2009; pp. 1670–1673.

19. Gaddam, V.R.; Yernagula, J.; Anantha, R.R.; Kona, S.; Kopparthi, S.; Chamakura, A.; Ajmera, P.K.; Srivastava, A. Remote power delivery for hybrid integrated bio-implantable electrical stimulation system. *Proc. SPIE* **2005**, *5763*, 20–31.

20. Li, P.F.; Bashirullah, R.; Principe, J.C. A low power battery management system for rechargeable wireless implantable electronics. In Proceedings of 2006 IEEE International Symposium on Circuits and Systems, 2006, ISCAS 2006, Island of Kos, Greece, 21–24 May 2006; pp. 1139–1142.

21. Evans, A.T.; Chiravuri, S.; Gianchandani, Y.B. Transdermal power transfer for recharging implanted drug delivery devices via the refill port. *Biomed. Microdevices* **2010**, *12*, 179–185.

22. Cong, P.; Chaimanonart, N.; Ko, W.H.; Young, D.J. A wireless and batteryless 10-bit implantable blood pressure sensing microsystem with adaptive RF powering for real-time laboratory mice monitoring. *IEEE J. Solid-State Circuits* **2009**, *44*, 3631–3644.

23. Cong, P.; Ko, W.H.; Young, D.J. Wireless implantable blood pressure sensing microsystem design for monitoring of small laboratory animals. *Sens. Mater.* **2008**, *20*, 327–340.

24. Cong, P.; Ko, W.H.; Young, D.J. Integrated electronic system design for an implantable wireless batteryless blood pressure sensing microsystem. *IEEE Commun. Mag.* **2010**, *48*, 98–104.

25. Cong, P.; Ko, W.H.; Young, D.J. Wireless batteryless implantable blood pressure monitoring microsystem for small laboratory animals. *IEEE Sens. J.* **2010**, *10*, 243–254.

26. Gosselin, B.; Simard, V.; Sawan, M. Low-power implantable microsystem intended to multichannel cortical recording. *IEEE Int. Symp. Circuits Syst.* **2004**, *4*, 5–8.

27. Martelli, D.; Silvani, A.; McAllen, R.M.; May, C.N.; Ramchandra, R. The low frequency power of heart rate variability is neither a measure of cardiac sympathetic tone nor of baroreflex sensitivity. *Am. J. Physiol. Heart. Circ. Physiol.* **2014**, *307*, doi:10.1152/ajpheart.00361.2014.

28. Huang, S.C.; Wong, M.K.; Lin, P.J.; Tsai, F.C.; Fu, T.C.; Wen, M.S.; Kuo, C.T.; Wang, J.S. Modified high-intensity interval training increases peak cardiac power output in patients with heart failure. *Eur. J. Appl. Physiol.* **2014**, *114*, 1853–1862.

29. Grodin, J.L.; Dupont, M.; Mullens, W.; Taylor, D.O.; Starling, R.C.; Tang, W. The prognostic role of cardiac power indices in advanced chronic heart failure. *J. Heart Lung Transpl.* **2014**, *33*, doi:10.1016/j.healun.2014.01.133.

30. Mun, J.Y.; Seo, M.G.; Kang, W.G.; Jun, H.Y.; Park, Y.H.; Pack, J.K. Study on the human effect of a wireless power transfer device at low frequency. In Proceedings of 2012 Progress in Electromagnetics Research Symposium (PIERS 2012), Moscow, Russia, 19–23 August 2012; pp. 322–324.

31. Leung, H.Y.; Budgett, D.M.; Taberner, A.; Hu, P. Power loss measurement of implantable wireless power transfer components using a peltier device balance calorimeter. *Meas. Sci. Technol.* **2014**, *25*, doi:10.1088/0957-0233/25/9/095010.

32. Shmilovitz, D.; Ozeri, S.; Wang, C.C.; Spivak, B. Noninvasive control of the power transferred to an implanted device by an ultrasonic transcutaneous energy transfer link. *IEEE Trans. BioMed. Eng.* **2014**, *61*, 995–1004.

33. Wang, J.X.; Wang, X.P.; Ma, Y.J.; Liu, N.; Yang, Z.Y. Analysis of heat transfer in power split device for hybrid electric vehicle using thermal network method. *Adv. Mech. Eng.* **2014**, *2014*, 210170.

34. Jonah, O.; Georgakopoulos, S.V.; Tentzeris, M.M. Wireless power transfer to mobile wearable device via resonance magnetic. In Proceedings of IEEE 14th Annual Wireless and Microwave Technology Conference (WAMICON), Orlando, FL, USA, 7–9 April 2013; pp. 1–3.

35. Krop, D.C.J.; Jansen, J.W.; Lomonova, E.A. Decoupled modeling in a multifrequency domain: Integration of actuation and power transfer in one device. *IEEE Trans. Magn.* **2013**, *49*, 3009–3019.

36. Van Mastrigt, R.; de Zeeuw, S.; Boeve, E.R.; Groen, J. Diagnostic power of the noninvasive condom catheter method in patients eligible for transurethral resection of the prostate. *Neurourol. Urodynam.* **2014**, *33*, 408–413.

37. Myers, K.A.; Leung, M.T.; Potts, M.T.; Potts, J.E.; Sandor, G.G.S. Noninvasive assessment of vascular function and hydraulic power and efficiency in pediatric fontan patients. *J. Am. Soc. Echocardiog.* **2013**, *26*, 1221–1227.

38. Suzuki, S.; Ishihara, M.; Kobayashi, Y. The improvement of the noninvasive power-supply system using magnetic coupling for medical implants. *IEEE Trans. Magn.* **2011**, *47*, 2811–2814.

39. Li, Q.B.; Liu, J.Q.; Zhang, G.J. Research on continuum power regression in noninvasive measurement of human blood glucose. *Spectrosc. Spect. Anal.* **2011**, *31*, 1481–1485.

40. Rodger, D.C.; Fong, A.J.; Wen, L.; Ameri, H.; Ahuja, A.K.; Gutierrez, C.; Lavrov, I.; Hui, Z.; Menon, P.R.; Meng, E.; *et al.* Flexible parylene-based multielectrode array technology for high-density neural stimulation and recording. *Sens. Actuat. B Chem.* **2008**, *132*, 449–460.

41. Wei, P.; Taylor, R.; Ding, Z.; Chung, C.; Abilez, O.J.; Higgs, G.; Pruitt, B.L.; Ziaie, B. Stretchable microelectrode array using room-temperature liquid alloy interconnects. *J. Micromech. Microeng.* **2011**, *21*, doi:10.1088/0960-1317/21/5/054015.

42. Rui, Y.F.; Liu, J.Q.; Yang, B.; Li, K.Y.; Yang, C.S. Parylene-based implantable platinum-black coated wire microelectrode for orbicularis oculi muscle electrical stimulation. *Biomed. Microdevices* **2012**, *14*, 367–373.

13

Simulation Study on Polarization-Independent Microlens Arrays Utilizing Blue Phase Liquid Crystals with Spatially-Distributed Kerr Constants

Hung-Shan Chen, Michael Chen, Chia-Ming Chang, Yu-Jen Wang and Yi-Hsin Lin *

Department of Photonics, National Chiao Tung University, Hsinchu 30010, Taiwan;
E-Mails: convince.eo98g@g2.nctu.edu.tw (H.-S.C.); michaelchen.eo01g@g2.nctu.edu.tw (M.C.);
spp02042003.eo02g@nctu.edu.tw (C.-M.C.); wangyujen.eo02g@g2.nctu.edu.tw (Y.-J.W.)

* Author to whom correspondence should be addressed; E-Mail: yilin@mail.nctu.edu.tw (Y.-H.L.)

External Editor: Hongrui Jiang

Abstract: Polarization independent liquid crystal (LC) microlens arrays based on controlling the spatial distribution of the Kerr constants of blue phase LC are simulated. Each sub-lens with a parabolic distribution of Kerr constants results in a parabolic phase profile when a homogeneous electric field is applied. We evaluate the phase distribution under different applied voltages, and the focusing properties of the microlens arrays are simulated. We also calculate polarization dependency of the microlenses arrays at oblique incidence of light. The impact of this study is to provide polarizer-free, electrically tunable focusing microlens arrays with simple electrode design based on the Kerr effect.

Keywords: Kerr effect; microlenses; polarization independent; liquid crystal; blue phase

1. Introduction

Liquid crystal (LC) microlens arrays are important in applications of 2D/3D switching, fiber coupling, and sensors [1–3]. Most of proposed structures of LC microlens arrays require at least one polarizer. To remove the usage of a polarizer, polarization independent LC phase modulations are developed. Three types of polarization independent LC phase modulations have been proposed: the type of the double-layered structure, the type of the residual phase structure, and the mixed type [4–10].

However, the structures were relatively complicated and the response times were slow. In 2010, we proposed a polarization independent polymer stabilized blue phase liquid crystal (PSBP-LC) microlens arrays based on the electric-field-induced Kerr effect, the field-induced birefringence is proportional to the electric field squared [11]. The Kerr effect exists in many LC materials, such as polymer stabilized isotropic phase liquid crystals, nematic liquid crystals, blue phase liquid crystals, and even ferroelectric liquid crystals [12–14]. In this paper, we proposed polarization independent LC microlens arrays based on controlling the distribution of the Kerr constants of blue phase LC (BPLC). The simulated results indicate the distribution of the Kerr constants of BPLC results in a parabolic optical phase shift and the proposed microlens arrays are capable of imaging. The polarization dependency of the LC microlens arrays is also discussed. The purpose of this study is mainly to provide a way to achieve polarizer-free, electrically tunable focusing microlens arrays with simple electrode design based on the Kerr effect.

2. Operating Principle and Lens Design

The Kerr medium, such as BPLC and PSBP-LC, is optically isotropic without an external electric field [11]. Under an external electric field (E), the optical axis of the field-induced birefringence is parallel to the electric field. The field-induced birefringence (Δn) is written as [15]:

$$\Delta n = n_e(E) - n_o(E) = \lambda \cdot K \cdot E^2 \tag{1}$$

where n_o is ordinary refractive index, n_e is extraordinary refractive index, K is the Kerr constant of the LC materials, and λ is wavelength of the incident light. Regarding the local orientations of LC molecules of BPLC under external electric field, $n_o(E)$ and $n_e(E)$ can be further expressed in Equations (2) and (3):

$$n_o(E) = n_{ave} - \frac{\lambda \cdot K \cdot E^2}{3} \tag{2}$$

$$n_e(E) = n_{ave} + \frac{2}{3} \cdot \lambda \cdot K \cdot E^2 \tag{3}$$

where n_{ave} represents the average refractive index without any applied electric field (i.e., $n_{ave} = (n_e + 2n_o)/3$). As a result, polarization independent phase modulation based on the Kerr effect of LC materials can be achieved. To generate a corresponding polarization independent phase profile of a lens, an inhomogeneous electric field is a way to be adopted [11]. However, the patterned electrodes are required. Instead of patterned electrodes, we proposed a spatially-distributed Kerr constant to achieve polarization independent microlens arrays. The structure and operating principles are depicted in Figure 1a,b. The structure primary consists of LC materials and two glass substrates coated with a layer of indium-tin-oxide (ITO). Without an applied voltage (V), an incident unpolarized light propagating along z-direction sees the average refractive index of n_{ave} because the effective optical index-ellipsoids are spherical which means the LC material is optically isotropic due to the cubic symmetry of the lattice structure, as depicted in Figure 1a [11,16]. With an applied voltage, an incident unpolarized light sees a spatial optical phase difference originating from a spatial distribution of Kerr constants, as depicted in Figure 1b. Assume the Kerr constant is spatially distributed in a parabolic form which can be expressed as:

$$K(r) = K_c - \frac{K_c - K_b}{r_0^2} \cdot r^2 \tag{4}$$

where r_0 is the radius of aperture of a sub-lens, r is position, K_c is the Kerr constant at the center of the aperture, and K_b is the Kerr constant around the peripheral region. Optical phase difference (OPD) under an applied voltage ($\delta(r)$) is $2\pi/\lambda \cdot [n_o(E) \cdot d]$, where d is the cell gap. From Equations (2) and (4), OPD is:

$$\delta(r) = \frac{2\pi}{\lambda} \times d \times \left[n_{ave} - \frac{\lambda \times E^2}{3} \times \left(K_c - \frac{K_c - K_b}{r_0^2} r^2 \right) \right] \tag{5}$$

r^2 term in Equation (4) is related to the focal length (f), inverse of lens power (P) [17,18]. Lens power is the degree that a lens converges or diverges light. The unit of lens power is diopter (D or m^{-1}). Thereafter, the lens power is written as:

$$P(E) = -\frac{2 \times \lambda \times d \times E^2 \times (K_c - K_b)}{3 \times r_0^2} = -\frac{2 \times \lambda \times d \times E^2 \times \Delta K}{3 \times r_0^2} \tag{6}$$

where ΔK is defined as $(K_c - K_b)$. Thus, we can realize microlens arrays based on spatially- distributed Kerr constants whose lens power is electrically tunable. The lens power of the mocrolens arrays is larger as both the applied electric field and ΔK are larger.

Figure 1. The structure and operating principles of the LC (liquid crystal) microlens arrays (**a**) without an applied voltage and (**b**) with an applied voltage.

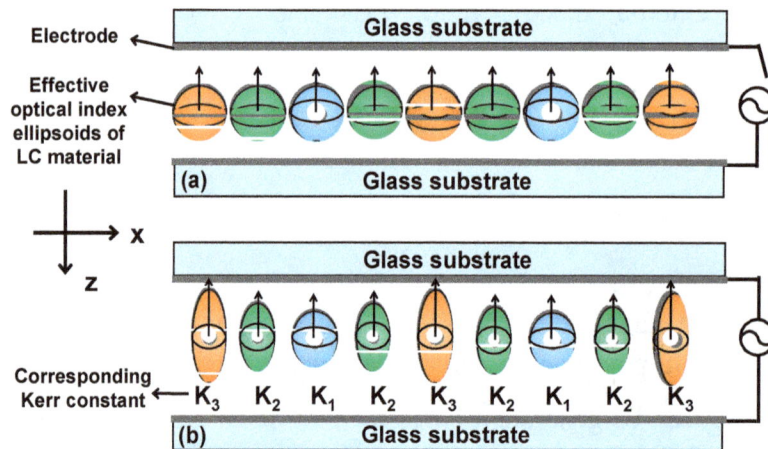

3. Simulation Results and Discussion

Here we simulate LC microlens arrays with spatially-distributed Kerr constants. The designed aperture size and the spacing between adjacent sub-lenses are 100 μm. The cell gap of the LC lenses is 25 μm. Usually Kerr constant is in a range between 10^{-8} and 10^{-10} V^2/m [19–21]. To demonstrate microlens arrays with a positive focal length, we design the Kerr constant in the center of a sub-lens (K_c) is 10^{-8} V^2/m while the Kerr constant at the peripheral region of a sub-lens (K_b) is 10^{-9} V^2/m. Figure 2 plots the parabolic distribution of Kerr constants of the microlens arrays based on the parameters we designed. We defined the phase shift as the difference between OPD at an applied voltage (V) and at $V = 0$. From Figure 2a and Equation (5), the phase shift as a function of position is shown in Figure 2b. The curve of phase shift in Figure 2b exhibits a periodically parabolic form at $V > 0$

due to the parabolic distribution of Kerr constant of the LC layer. The phase shift increases with an applied voltage. Based on Equation (6) and the parabolic distribution we designed, the simulated voltage-dependent lens power is depicted in Figure 2c. The lens power increases with an applied voltage. The lens power at $V = 100$ V_{rms} is around 650 m^{-1}, which is corresponding to the focal length of ~1.54 mm.

Figure 2. (**a**) The spatial distribution of Kerr constants of the LC microlens arrays based on the parameters we designed; (**b**) the corresponding spatial phase shift of the LC microlens arrays at $V = 0$ (red line), 50 V_{rms} (green line), and 100 V_{rms} (blue line); and (**c**) the simulated voltage- dependent lens power.

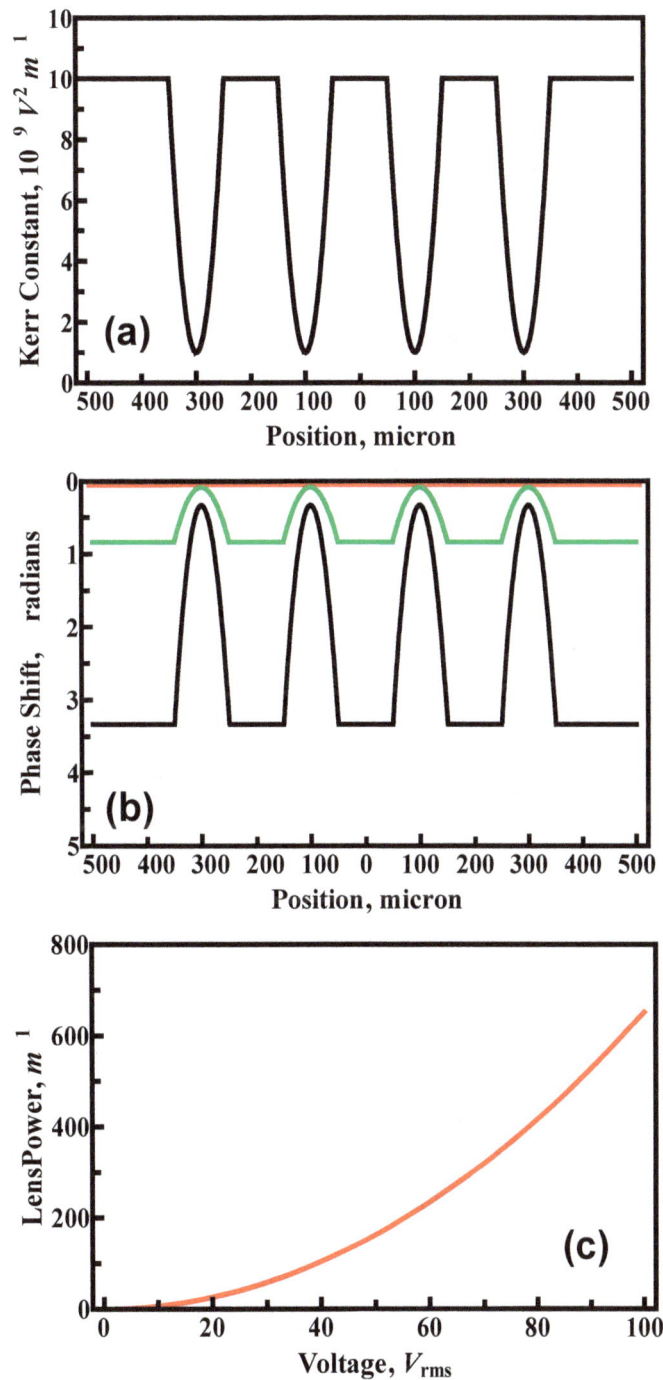

To simulate the focusing properties of the microlens arrays at the focal plane, we adopted the Fresnel approximation [22]. Figure 3a,b shows the spatial phase shift and corresponding intensity distribution at the focal plane of the micolens arrays. As we can see, the parabolic phase shift (red line in Figure 3a results in sharp peaks at the focal plane (red line in Figure 3b). In contrast, the trapezoid-like phase shift (blue dotted line in Figure 3a) results in relatively broad peaks at the focal plane (blue dotted line in Figure 3b). Therefore, the parabolic phase distribution is necessary to realize good imaging quality which also means the distribution of Kerr constants should be parabolic. In 2011, Wu *et al.* proposed an Eiffel-Tower-like ITO electrode to generate an ideal phase distribution in BPLC [23]. However, the Eiffel-Tower-like ITO electrode is difficult to fabricate. The method of the spatial distribution of Kerr constants that we proposed is more practical because our method does not require complex electrodes.

Figure 3. (**a**) The simulated spatial phase shifts of the LC microlens arrays. Blue dotted line stands for periodically trapezoid-like phase shift and red line stands for periodically parabolic phase shift; (**b**) the corresponding intensity distribution at the focal plane for the periodically trapezoid-like phase shift (dotted blue line) and the periodically parabolic phase shift (red line).

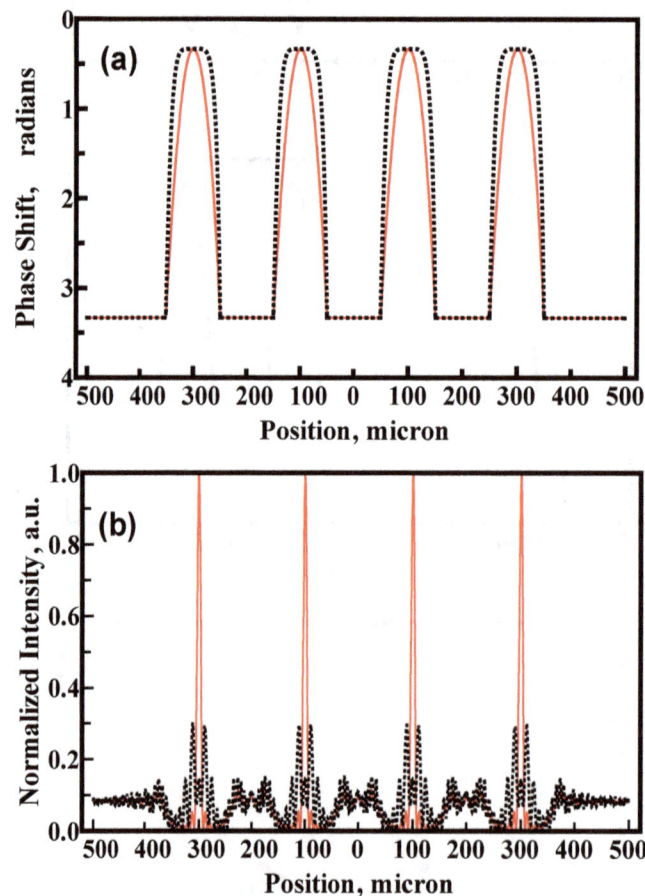

The LC microlens arrays as incident light is at the oblique angle (*i.e.*, off-axis) is also important in applications. Assume the incident angle is θ_i with respect to *z*-direction and the light propagates in LC cell with an angle of θ_{LC}. Because the change of the refractive index of the Kerr medium is very small (normally < 0.05), we can assume the incident light propagates in a straight way in the medium and θ_{LC}

is able to be deduced from Snell's law (*i.e.*, $n_{air} \times \sin\theta_i = n_{ave} \times \sin\theta_{LC}$). Two eigenmodes propagating in the LC medium are defined as e-mode and the o-mode. The polarization of e-mode lies in the plane of *x-z* plane and that of o-mode is perpendicular to *x-z* plane. Thereafter, we can calculate the phase shift of the e-mode as:

$$\delta_{e-mode}(E,x) = \frac{2\pi}{\lambda}\int n_{e,eff}(E,\theta_{LC},x)\cdot dk = \frac{2\pi}{\lambda}\int n_{e,eff}(E,\theta_{LC},x)\cdot\csc\theta_{LC}\cdot dx \tag{7}$$

where the effective extraordinary refractive index ($n_{e,eff}$) can be expressed as:

$$n_{e,eff}(E,\theta_{LC},x) = (\frac{n_e^2(E,x)}{\sin^2(\theta_{LC})} + \frac{n_o^2(E,x)}{\cos^2(\theta_{LC})})^{-0.5} \tag{8}$$

The phase shift of the o-mode can also be expressed as:

$$\delta_{o-mode}(E,x) = \frac{2\pi}{\lambda}\int n_o(E,x)\cdot dk = \frac{2\pi}{\lambda}\int n_o(E,x)\cdot\csc\theta_{LC}\cdot dx \tag{9}$$

To simplify the discussion, a single sub-lens is considered in the following discussions. The diameter of the sub-lens and the cell gap are 100 and 25 μm, respectively. The simulated phase shifts as incident light is at the oblique angle are shown in Figure 4. The dotted line represents the phase shift of e-mode while the solid line represents the o-mode at $V = 100$ V$_{rms}$. The blue and the red represent θ_{LC} of ~+10° and −10°, respectively. From Snell's law, θ_i is ~±15.7° when corresponding θ_{LC} is ±10° and n_{ave} is around 1.56. From the simulation results, the phase shift between the center and the peripheral region for the e-mode is around 3π radians. As to the o-mode, the phase shift between the center and the peripheral region is around ~2.8π. This also indicates the micorlens arrays are polarization dependent at the oblique incidence because the refractive index changes more for o-ray than that for e-ray. To reduce the polarization dependency of the phase shift at oblique incidence, we can use other electrically tunable LC cells for phase compensation.

To experimentally realize the spatial distribution of the Kerr constants of the LC materials, one can produce the spatial distribution of Kerr constants of the Kerr medium, such as BPLC, in terms of fabrication method of spatial temperature gradient. Based on previous research results, the Kerr constant is proportional to the coherent length squared (ξ^2), inversely proportional to $T-T^*$, where T is temperature and T^* represents the temperature as the coherent length of the LC become infinite (*i.e.*, $K \propto \xi^2 \propto 1/T - T^*$) [19–21]. As a result, the Kerr constant of BPLC strongly depends on the temperature. Therefore, the spatial distribution of the Kerr constant can be controlled by means of temperature gradient, and then we can use photo-polymerization to stabilize BPLC in order to regulate the distribution of phase separation and further to generate spatially-distributed Kerr constants. To demonstrate the proposed idea, we step-controlled the curing temperatures of the PSBP-LC materials, and we realized the Kerr constant difference check by phase retardation measurement three times. However, due to the limit of the temperature gradient controlling instrument, we are not able to put such a big Kerr constant difference within this small aperture region. For further implementation of this concept, one might need step masks or more precision thermal controlling machines to get steep phase distribution within the aperture.

Figure 4. The simulated phase shift at the oblique angle. The dotted line represents the phase shift of o-mode and the solid line represents the phase shift of e-mode. The blue and red represent the incident angle θ_i is +15.7° and −15.7°, respectively.

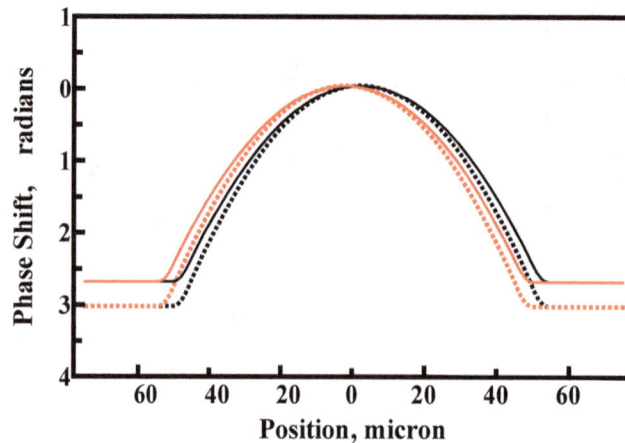

4. Conclusions

We proposed a polarization independent LC microlens arrays based on spatially distributed Kerr constants of the LC material. The mechanism and simulated performance are discussed. In addition, we also evaluate the polarization dependency of the microlens arrays at oblique angle of incidence. This study provides a method to realize polarizer-free and electrically tunable microlens arrays with simple electrodes based on the Kerr effect.

Acknowledgments

This research was supported partially by the Ministry of Science and Technology (MOST) in Taiwan under the contract No. NSC 101-2112-M-009-011-MY3.

Author Contributions

Hung-Shan Chen proposed the concept, did the simulation and wrote the manuscript. Michael Chen, Chia-Ming Chang, and Yu-Jen Wang helped in the simulations and discussions. Yi-Hsin Lin discussed the concept and results with Hung-Shan Chen, and also revised the manuscript.

Conflicts of Interest

The authors declare no conflict of interest

References

1. Liu, Y.F.; Ren, H.W.; Xu, S.; Li, Y.; Wu, S.T. Fast-response liquid crystal lens for 3D displays. *SPIE Proc.* **2014**, *9005*, 1–10.
2. Chen, M.; Chen, C.H.; Lai, Y.; Lu, Y.C.; Lin, Y.H. An electrically tunable polarizer for a fiber system based on a polarization-dependent beam size derived from a liquid crystal lens. *IEEE Photonics J.* **2014**, *6*, 1–8.

3. Klaus, W.; Ide, M.; Hayano, Y.; Morokawa, S.; Arimoto, Y. Adaptive LC lens array and its application. *SPIE Proc.* **1999**, *3635*, 66–73.

4. Lin, Y.H.; Ren, H.W.; Wu, Y.H.; Zhao, Y.; Fang, J.Y.; Ge, B.Z.; Wu, S.H. Polarization-independent liquid crystal phase modulator using a thin polymer-separated double-layered structure. *Opt. Express* **2005**, *13*, 8746–8752.

5. Lin, Y.H.; Ren, H.W.; Fan, Y.H.; Wu, Y.H.; Wu, S.T. Polarization-independent and fast-response phase modulation using a normal-mode polymer-stabilized cholesteric texture. *J. Appl. Phys. Lett.* **2005**, *98*, doi:10.1063/1.2037191.

6. Ren, H.W.; Lin, Y.H.; Fan, Y.H.; Wu, S.T. Polarization-independent phase modulation using a polymer-dispersed liquid crystal. *Appl. Phys. Lett.* **2005**, *86*, doi:10.1063/1.1899749.

7. Wu, Y.H.; Lin, Y.H.; Lu, Y.Q.; Ren, H.; Fan, Y.H.; Wu, J.R.; Wu, S.T. Submillisecond response variable optical attenuator based on sheared polymer network liquid crystal. *Opt. Express* **2004**, *12*, 6382–6389.

8. West, J.L.; Zhang, G.Q.; Reznikov, Y.; Glushchenko, A. Fast birefringent mode of stressed liquid crystal. *Appl. Phys. Lett.* **2005**, *86*, doi:10.1063/1.1852720.

9. Ren, H.; Lin, Y.H.; Wen, C.H.; Wu S.T. Polarization-independent phase modulation of a homeotropic liquid crystal gel. *Appl. Phys. Lett.* **2005**, *87*, doi:10.1063/1.2126107.

10. Lin, T.H.; Chen, M.S.; Lin, W.C.; Tsou, Y.S. A polarization-independent liquid crystal phase modulation using polymer-network liquid crystals in a 90° twisted cell. *J. Appl. Phys. Lett.* **2012**, *112*, doi:10.1063/1.4737260.

11. Lin, Y.H.; Chen, H.S.; Lin, H.C.; Tsou, Y.S.; Hsu, H.K.; Li, W.Y. Polarizer-free and fast response microlens arrays using polymer-stabilized blue phase liquid crystals. *Appl. Phys. Lett.* **2010**, *96*, doi:10.1063/1.3360860.

12. Yang, Y.C.; Yang, D.K. Electro-optic Kerr effect in polymer-stabilized isotropic liquid crystals. *Appl. Phys. Lett.* **2011**, *98*, doi:10.1063/1.3533396.

13. Khoshsima, H.; Tajalli, H.; Ghanadzadeh Gilani, A.; Dabrowski, R. Electro-optical Kerr effect of two high birefringence nematic liquid crystals. *Appl. Phys. Lett.* **2006**, *39*, 1495–1499.

14. Pozhidaev, E.P.; Kiselev, A.D.; Srivastave, A.K.; Chigrinov, V.G.; Kwok, H.S.; Minchenko, M.V. Orientational "Kerr effect" and phase modulaion of light in deformed-helix ferroelectric liquid crystals with subwavelength pitch. *Phys. Rev.* **2013**, *87*, doi:10.1103/PhysRevE.87.052502.

15. Lin, Y.H.; Chen, H.S.; Wu, C.H.; Hsu, H.K. Measuring electric-field-induced birefringence in polymer stabilized blue phase liquid crystal based on phase shift measurement. *J. Appl. Phys.* **2011**, *109*, doi:10.1063/1.3583572.

16. Amnon, Y.; Yeh, P. *Optical Waves in Crystals: Propagation and Control of Laser Radiation*; Wiley: New York, NY, USA, 1984; p. 83.

17. Lin, H.C.; Chen, M.S.; Lin, Y.H. A review of electrically tunable focusing liquid crystal lenses. *Trans. Electr. Electron. Mater.* **2011**, *12*, 234–240.

18. Lin, Y.H.; Chen, H.S. Electrically tunable-focusing and polarizer-free liquid crystal lenses for ophthalmic applications. *Opt. Express* **2013**, *21*, 9428–9436.

19. Haseba, Y.; Kikuchi, H.; Nagamura, T.; Kajiyama, T. Large electr-optic Kerr effect in nanostructures chiral liquid-crystal composites over a wide temperature range. *Adv. Mater.* **2005**, *17*, 2311–2315.

20. Tian, L.; Goodby, J.W.; Gortz, V.; Gleeson, H.F. The magnitude and tmeperature dependence of the Kerr constant in liquid crystal blue phases and the dark conglomerate phase. *Liq. Crys.* **2013**, *40*, 1446–1454.

21. Majles Ara, M.H.; Mousavi, S.H.; Rafiee, M.; Zakerhamidi, M.S. Dtermination of temperature dependence of Kerr constant for nematic liquid crystal. *Mol. Cryst. Liq. Cryst.* **2011**, *544*, 227/[1215]–231/[1219]

22. Goodman, J.W. *Introduction to Fourier Optics*, 3rd ed.; Roberts and Company Publishers: Greenwood Village, CO, USA, 2005; p. 67.

23. Li, Y.; Wu, S.T. Polarization independent adaptive microlens with a blue-phase liquid crystal. *Opt. Express* **2011**, *19*, 8045–8050.

Permissions

The contributors of this book come from diverse backgrounds, making this book a truly international effort. This book will bring forth new frontiers with its revolutionizing research information and detailed analysis of the nascent developments around the world.

We would like to thank all the contributing authors for lending their expertise to make the book truly unique. They have played a crucial role in the development of this book. Without their invaluable contributions this book wouldn't have been possible. They have made vital efforts to compile up to date information on the varied aspects of this subject to make this book a valuable addition to the collection of many professionals and students.

This book was conceptualized with the vision of imparting up-to-date information and advanced data in this field. To ensure the same, a matchless editorial board was set up. Every individual on the board went through rigorous rounds of assessment to prove their worth. After which they invested a large part of their time researching and compiling the most relevant data for our readers.

The editorial board has been involved in producing this book since its inception. They have spent rigorous hours researching and exploring the diverse topics which have resulted in the successful publishing of this book. They have passed on their knowledge of decades through this book. To expedite this challenging task, the publisher supported the team at every step. A small team of assistant editors was also appointed to further simplify the editing procedure and attain best results for the readers.

Apart from the editorial board, the designing team has also invested a significant amount of their time in understanding the subject and creating the most relevant covers. They scrutinized every image to scout for the most suitable representation of the subject and create an appropriate cover for the book.

The publishing team has been an ardent support to the editorial, designing and production team. Their endless efforts to recruit the best for this project, has resulted in the accomplishment of this book. They are a veteran in the field of academics and their pool of knowledge is as vast as their experience in printing. Their expertise and guidance has proved useful at every step. Their uncompromising quality standards have made this book an exceptional effort. Their encouragement from time to time has been an inspiration for everyone.

The publisher and the editorial board hope that this book will prove to be a valuable piece of knowledge for researchers, students, practitioners and scholars across the globe.

List of Contributors

Tao Luo
Key Laboratory of Instrumentation Science and Dynamic Measurement (North University of China),
Ministry of Education, North University of China, Taiyuan 030051, China

Qiulin Tan
Key Laboratory of Instrumentation Science and Dynamic Measurement (North University of China), Ministry of Education, North University of China, Taiyuan 030051, China
Science and Technology on Electronic Test and Measurement Laboratory, North University of China, Taiyuan 030051, China
National Key Laboratory of Fundamental Science of Micro/Nano-Device and System Technology, Chongqing University, Chongqing 400044, China

Liqiong Ding
Key Laboratory of Instrumentation Science and Dynamic Measurement (North University of China), Ministry of Education, North University of China, Taiyuan 030051, China

Tanyong Wei
Science and Technology on Electronic Test and Measurement Laboratory, North University of China, Taiyuan 030051, China

Chao Li
Science and Technology on Electronic Test and Measurement Laboratory, North University of China, Taiyuan 030051, China

Chenyang Xue
Key Laboratory of Instrumentation Science and Dynamic Measurement (North University of China), Ministry of Education, North University of China, Taiyuan 030051, China
Science and Technology on Electronic Test and Measurement Laboratory, North University of China, Taiyuan 030051, China

Jijun Xiong
Key Laboratory of Instrumentation Science and Dynamic Measurement (North University of China), Ministry of Education, North University of China, Taiyuan 030051, China
Science and Technology on Electronic Test and Measurement Laboratory, North University of China, Taiyuan 030051, China

Andrea Masiero
CIRGEO (Interdepartmental Research Center of Geomatics), University of Padova, via dell'Università 16, 35020 Legnaro (PD), Italy

Alberto Guarnieri
CIRGEO (Interdepartmental Research Center of Geomatics), University of Padova, via dell'Università 16, 35020 Legnaro (PD), Italy

Francesco Pirotti
CIRGEO (Interdepartmental Research Center of Geomatics), University of Padova, via dell'Università 16, 35020 Legnaro (PD), Italy

Antonio Vettore
CIRGEO (Interdepartmental Research Center of Geomatics), University of Padova, via dell'Università 16, 35020 Legnaro (PD), Italy

Dimitry Dumont-Fillon
Debiotech SA, 28 avenue de Sévelin, 1004 Lausanne, Switzerland

Hassen Tahriou
Debiotech SA, 28 avenue de Sévelin, 1004 Lausanne, Switzerland

Christophe Conan
Debiotech SA, 28 avenue de Sévelin, 1004 Lausanne, Switzerland

Eric Chappel
Debiotech SA, 28 avenue de Sévelin, 1004 Lausanne, Switzerland

Slawomir Jakiela
Institute of Physical Chemistry, Polish Academy of Sciences, Kasprzaka 44/52, 01-224 Warsaw, Poland

Pawel R. Debski
Institute of Physical Chemistry, Polish Academy of Sciences, Kasprzaka 44/52, 01-224 Warsaw, Poland

Bogdan Dabrowski
Institute of Physical Chemistry, Polish Academy of Sciences, Kasprzaka 44/52, 01-224 Warsaw, Poland

Piotr Garstecki
Institute of Physical Chemistry, Polish Academy of Sciences, Kasprzaka 44/52, 01-224 Warsaw, Poland

Anna Haller
Institute of Sensor and Actuator Systems, Vienna University of Technology, Gusshausstrasse 27-29/E366, 1040 Vienna, Austria

Andreas Spittler
Core Facility Flow Cytometry & Department of Surgery, Research Laboratories, Center of Translational Research, Medical University of Vienna, Lazarettgasse 14, 1090 Vienna, Austria

Lukas Brandhoff
Institute of Microsensors, -Actuators and -Systems (IMSAS) & Microsystems Center Bremen (MCB), University of Bremen, Otto-Hahn-Allee 1, 28359 Bremen, Germany

Helene Zirath
Health and Environment Department, Austrian Institute of Technology, Muthgasse 11, 1190 Vienna, Austria

Dietmar Puchberger-Enengl
Institute of Sensor and Actuator Systems, Vienna University of Technology, Gusshausstrasse 27-29/E366, 1040 Vienna, Austria

Franz Keplinger
Institute of Sensor and Actuator Systems, Vienna University of Technology, Gusshausstrasse 27-29/E366, 1040 Vienna, Austria

Michael J. Vellekoop
Institute of Microsensors, -Actuators and -Systems (IMSAS) & Microsystems Center Bremen (MCB), University of Bremen, Otto-Hahn-Allee 1, 28359 Bremen, Germany

Miao Xu
BK Plus Haptic Polymer Composite Research Team, Department of Polymer-Nano Science and Technology, Chonbuk National University, Jeonju, Chonbuk 561-756, Korea

Xiahui Wang
BK Plus Haptic Polymer Composite Research Team, Department of Polymer-Nano Science and Technology, Chonbuk National University, Jeonju, Chonbuk 561-756, Korea

Boya Jin
BK Plus Haptic Polymer Composite Research Team, Department of Polymer-Nano Science and Technology, Chonbuk National University, Jeonju, Chonbuk 561-756, Korea

Hongwen Ren
BK Plus Haptic Polymer Composite Research Team, Department of Polymer-Nano Science and Technology, Chonbuk National University, Jeonju, Chonbuk 561-756, Korea

Ravi Prakash
Biosystems Research and Applications Group, Department of Electrical and Computer Engineering, Schulich School of Engineering, University of Calgary, Calgary, AB T2N 1N4, Canada

Kanti Pabbaraju
Provincial Laboratory for Public Health of Alberta, Calgary, AB T2N 4W4, Canada

Sallene Wong
Provincial Laboratory for Public Health of Alberta, Calgary, AB T2N 4W4, Canada

Anita Wong
Provincial Laboratory for Public Health of Alberta, Calgary, AB T2N 4W4, Canada

Raymond Tellier
Provincial Laboratory for Public Health of Alberta, Calgary, AB T2N 4W4, Canada
Department of Microbiology, Immunology and Infectious Diseases, Cumming School of Medicine, University of Calgary, Calgary, AB T2N 1N4, Canada

Karan V. I. S. Kaler
Biosystems Research and Applications Group, Department of Electrical and Computer Engineering, Schulich School of Engineering, University of Calgary, Calgary, AB T2N 1N4, Canada

Sasha Cai Lesher-Perez
Department of Biomedical Engineering, Biointerfaces Institute, University of Michigan, Ann Arbor, MI 48109, USA

Priyan Weerappuli
Department of Biomedical Engineering, Biointerfaces Institute, University of Michigan, Ann Arbor, MI 48109, USA
Department of Biomedical Engineering, Wayne State University, Detroit, MI 48202, USA
Department of Physiology, Wayne State University, Detroit, MI 48201, USA

Sung-Jin Kim
Department of Biomedical Engineering, Biointerfaces Institute, University of Michigan, Ann Arbor, MI 48109, USA
Department of Mechanical Engineering, Konkuk University, Seoul 143-701, Korea Chongqing 400030, China

Chao Zhang
Department of Biomedical Engineering, Biointerfaces
Institute, University of Michigan, Ann Arbor, MI 48109,
USA
Key Laboratory of Low-Grade Energy Utilization
Technologies and Systems, Chongqing University,
Chongqing 400030, China
Institute of Engineering Thermophysics, Chongqing
University, Chongqing 400030, China

Shuichi Takayama
Department of Biomedical Engineering, Biointerfaces
Institute, University of Michigan, Ann Arbor, MI 48109,
USA
Department of Macromolecular Science and Engineering,
University of Michigan, Ann Arbor, MI 48109, USA
Division of Nano-Bio and Chemical Engineering World
Class University Project, Ulsan National Institute of
Science and Technology, Ulsan 689-798, Korea

Sebastian M. Bonk
Chair of Biophysics, University of Rostock, Rostock
18057, Germany

Paul Oldorf
SLV Mecklenburg-Vorpommern GmbH, Rostock 18069,
Germany

Rigo Peters
SLV Mecklenburg-Vorpommern GmbH, Rostock 18069,
Germany

Werner Baumann
Chair of Biophysics, University of Rostock, Rostock
18057, Germany

Jan Gimsa
Chair of Biophysics, University of Rostock, Rostock
18057, Germany

Shifei Liu
College of Automation, Harbin Engineering University,
145 Nantong St., Nangang District, Harbin 150001, China

Mohamed Maher Atia
Department of Electrical and Computer Engineering,
Royal Military College of Canada, P.O. Box 17000, Station
Forces, Kingston, ON K7K 7B4, Canada

Yanbin Gao
College of Automation, Harbin Engineering University,
145 Nantong St., Nangang District, Harbin 150001, China

Aboelmagd Noureldin
Department of Electrical and Computer Engineering,
Royal Military College of Canada, P.O. Box 17000, Station
Forces, Kingston, ON K7K 7B4, Canada

Juliano Machado
Assistive Technology Laboratory, Federal Institute of
Rio Grande do Sul (IFSul), General Balbão Street 81,
Charqueadas 96745-000, Brazil

Biomedical Instrumentation Laboratory, Federal
University of Rio Grande do Sul (UFRGS), Avenue
Osvaldo Aranha 103, Porto Alegre 90035-190, Brazil

Alexandre Balbinot
Biomedical Instrumentation Laboratory, Federal
University of Rio Grande do Sul (UFRGS), Avenue
Osvaldo Aranha 103, Porto Alegre 90035-190, Brazil

Jing-Quan Liu
National Key Laboratory of Science and Technology on
Micro/Nano Fabrication, Key Laboratory of Shanghai
Education Commission for Intelligent Interaction and
Cognitive Engineering, Department of Micro/Nano-
electronics, Shanghai Jiao Tong University, Shanghai
200240, China

Yue-Feng Rui
National Key Laboratory of Science and Technology on
Micro/Nano Fabrication, Key Laboratory of Shanghai
Education Commission for Intelligent Interaction and
Cognitive Engineering, Department of Micro/Nano-
electronics, Shanghai Jiao Tong University, Shanghai
200240, China

Xiao-Yang Kang
National Key Laboratory of Science and Technology on
Micro/Nano Fabrication, Key Laboratory of Shanghai
Education Commission for Intelligent Interaction and
Cognitive Engineering, Department of Micro/Nano-
electronics, Shanghai Jiao Tong University, Shanghai
200240, China

Bin Yang
National Key Laboratory of Science and Technology on
Micro/Nano Fabrication, Key Laboratory of Shanghai
Education Commission for Intelligent Interaction and
Cognitive Engineering, Department of Micro/Nano-
electronics, Shanghai Jiao Tong University, Shanghai
200240, China

Xiang Chen
National Key Laboratory of Science and Technology on
Micro/Nano Fabrication, Key Laboratory of Shanghai
Education Commission for Intelligent Interaction and
Cognitive Engineering, Department of Micro/Nano-
electronics, Shanghai Jiao Tong University, Shanghai
200240, China

Chun-Sheng Yang
National Key Laboratory of Science and Technology on Micro/Nano Fabrication, Key Laboratory of Shanghai Education Commission for Intelligent Interaction and Cognitive Engineering, Department of Micro/Nano-electronics, Shanghai Jiao Tong University, Shanghai 200240, China

Hung-Shan Chen
Department of Photonics, National Chiao Tung University, Hsinchu 30010, Taiwan

Michael Chen
Department of Photonics, National Chiao Tung University, Hsinchu 30010, Taiwan

Chia-Ming Chang
Department of Photonics, National Chiao Tung University, Hsinchu 30010, Taiwan

Yu-Jen Wang
Department of Photonics, National Chiao Tung University, Hsinchu 30010, Taiwan

Yi-Hsin Lin
Department of Photonics, National Chiao Tung University, Hsinchu 30010, Taiwan

www.ingramcontent.com/pod-product-compliance
Lightning Source LLC
Chambersburg PA
CBHW070155240326
41458CB00126B/5169